Hydrogen
Medicine

Hydrogen Medicine

Combining Oxygen, Hydrogen, and CO2

DR. MARK SIRCUS

HYDROGEN MEDICINE
COMBINING OXYGEN, HYDROGEN, AND CO2

iUniverse books may be ordered through booksellers or by contacting:

iUniverse
1663 Liberty Drive
Bloomington, IN 47403
www.iuniverse.com
844-349-9409

ISBN: 978-1-6632-2350-0 (sc)
ISBN: 978-1-6632-2352-4 (hc)
ISBN: 978-1-6632-2351-7 (e)

Library of Congress Control Number: 2021911068

Print information available on the last page.

iUniverse rev. date: 06/15/2021

Contents

Preface

This book will explore hydrogen gas, hydrogen water, oxygen (O2), and carbon dioxide (CO2). Combining these gases will usher in a new age of medicine where the impossible becomes possible. Expect every protocol to perform better when the levels of these primordial gasses are optimized. Looking for the fountain of youth (anti-aging medicines), we find nothing as powerful as these gases. Everything done in ICU departments is safer when these gases are employed.

For most doctors, nurses, and patients, this is the first comprehensive look at hydrogen medicine. The sun loves hydrogen, so does water, and doctors will too because hydrogen offers an extraordinary safe way to treat people. Hydrogen is the most abundant element in the universe, and as a medicine, it stands up tall, though small, next to oxygen.

I recently received a most extraordinary testimonial about a patient with Multiple System Atrophy (MSA), an exceedingly rare neuro-degenerative progressively fatal disease, similar in many ways to ALS/ Motor Neuron Disease. The disease had already progressed for eight years in this patient. They usually would be deteriorating rapidly to the point of needing round-the-clock care with death on the horizon. Then this person started hydrogen inhalation therapy. After eight months of one of the safest treatments in medicine, instead of being on or near a death bed, the hydrogen gas brought this person back to health to full mobility and active life.

Hydrogen is known to save lives in ICU and emergency rooms. In China, doctors have already used hydrogen gas to address COVID-19 to address severe lung distress quickly. In this book, some stories lead one to believe that the worse a person's condition is, the better hydrogen works. Hydrogen works like a miracle gas when one is under tremendous pressure. Down at 2000 feet, deep-sea divers can survive at bone-crushing depths by breathing 96 percent hydrogen and only 4 percent oxygen. It is also used to prevent decompression and nitrogen sickness.

H_2 has shown anti-inflammatory and antioxidant ability in many clinical trials, and its application in the latest Chinese novel coronavirus pneumonia (NCP) treatment guidelines have been favorable. Clinical experiments have revealed the surprising finding that H_2 gas may protect the lungs and extrapulmonary organs from pathological stimuli in NCP patients.

The Seventh Edition of Chinese Clinical Guidance for COVID-19 Pneumonia Diagnosis and Treatment (7th Edition), issued by the Chinese National Health Commission, recommended the inhalation of oxygen mixed with hydrogen gas (33.3% O_2 and 66.6% H_2), bringing H_2 to the forefront of contemporary therapeutic medical gas research.

Molecular hydrogen (H_2) gas regulates anti-inflammatory and antioxidant activity, mitochondrial energy metabolism, endoplasmic reticulum stress, the immune system, and cell death (apoptosis, autophagy, pyroptosis, ferroptosis, and the circadian clock, among others) and has therapeutic potential for many systemic diseases

Molecular hydrogen penetrates deep into every cell of your body, fighting free radicals and harmful inflammation at the cellular level. H_2 is the lightest chemical element in the Earth's atmosphere.

Some of the research-backed benefits of molecular hydrogen include:

- Supports cognitive health
- Supports healthy immune function
- Supports healthy energy levels

Over 1400 peer-reviewed studies from across the globe have demonstrated that molecular hydrogen is a unique antioxidant. Some of the principal characteristics of hydrogen include:

Selectivity - H2 is such a stable molecule that it only reacts with the very worst free radicals: hydroxyl radical and peroxynitrite. This means it leaves the beneficial radicals used in cell signaling and immune function alone and only targets the "bad guys."

Size - Hydrogen is the smallest element. This allows it to diffuse very freely, allowing the free-radical fighting hydrogen molecules to get places most other antioxidants cannot. Hydrogen can cross the blood-brain barrier and cell membranes, penetrating a cell's nucleus and mitochondria.

Cell Signaling - Much of hydrogen's benefits come from its cell signaling activity. Hydrogen works at the genetic level within a cell by "turning on" the genes that code for our natural antioxidant defense systems. H2 tells your cells to start pumping out more of your natural antioxidants, like glutathione peroxidase (GSH) and superoxide dismutase (SOD).

Hydrogen Got Me Out of Bed Testimony.

Dear Dr. Sircus,

I was not doing the AquaCure for some time and even didn't realize how far I was declining again until it was difficult to get out of bed.

The other day I was completely bedridden. During prayer, I was reminded to use the machine, which was stored away. I breathed Hydrox for 20 minutes and drank the hydrogen-infused water. One hour later, I was up and could do my most important chores.

I was thankful all the rest of the day for such life-sustaining help.

I want so much to thank you, Dr. Sircus, for being a profound, important, critically needed teacher and healer in my life.

With my deepest gratitude, respect, and appreciation.
Josef

It is a beautiful feeling to receive testimonies like this. Knowing that one is helping others to feel and be better makes life worth living. My personal testimony is more about anti-aging. I look and feel more than ten years younger since beginning to use hydrogen inhalation therapy. Hydrogen is also the only thing that I have found that has almost completely reversed the neuropathy in my feet!

Hydrogen is the Backbone of the Universe.

Out of nothing comes something. Hydrogen is that something. Vast clouds of hydrogen occupy the immense reaches of space. Hydrogen fills every cubic meter of the universe. The process of life, photosynthesis, is all about attaching hydrogen to carbohydrates, fats, proteins.

H2 is a flammable, colorless, odorless gas. It was previously considered physiologically inert in mammalian cells and not to react with active substrates in biological systems. Now, H2 has emerged as a novel medical gas with broad applications.

The main point of this book is to promote hydrogen therapy. Anyone with a life-threatening disease needs hydrogen inhalation therapy to increase their security that their treatments will work no matter what other treatments are employed. This book is about leaping tall medical buildings in a single bound with the three primary medical gases.

Hydrogen Medicine
is Revolutionary.

When studying *Hydrogen Medicine,* one is embarking on a medical journey through the very basics of life. As the subtitle suggests, command of the three elemental gases will get one everywhere, medically speaking, when combined with essential minerals, detoxification protocols, organic diets, and intermittent fasting.

Hydrogen Medicine is revolutionary. In the future that this book envisions, oxygen will no longer be given alone but always with hydrogen.

It is time for the world to embrace hydrogen as a clean energy carrier to save us from the pollution nightmare affecting every urban center globally. It is also time for modern medicine to wake up to hydrogen's power and purity as a medicine. Hydrogen produces zero toxic emissions when used for energy, and it has zero side effects when used as a medicine.

Healing Power When We Need It.

When we are under tremendous stress, sick, or dying, a hydrogen inhaling machine should be right by our bed, sitting under our desk at the office, or both. Think of a deep-sea diver down 2,000 feet— incredible stress and pressure. To stay alive down there, divers breathe 96 percent hydrogen and only 4 percent oxygen.

At the death door, hydrogen's effects are most noticeable. That is why it belongs in ambulances, emergency centers, and Innovative ICU Medicine (Intensive Care) right alongside oxygen.

Hydrogen molecules and ions are the backbones of the universe; we will still see hydrogen running the sun and every other star in the universe in a thousand years. Hopefully, it will not take that long before hydrogen therapy is introduced during surgery, in clinics, spas, and in all homes.

Molecular Hydrogen is the Perfect Medical Treatment for Oxidative Stress.

Oxidative stress, caused by rivers of free radicals, is a plague on modern man. It is the toxic pollution, chemical exposures, heavy metals, radiation from the indiscriminate use of medical imaging, pharmaceutical medicines, chemotherapy, your cell phone, which increases oxidative stress.

From a cell's perspective, hydrogen inhalation is like standing under a refreshing waterfall. Inhaling hydrogen gas (with oxygen included) will extinguish oxidative stress and inflammation, just like a firetruck puts out fires.

Hydrogen therapy is a new, innovative clinical mode of treatment for many medical situations, including surgery, tissue damage and dysfunction, diabetes, heart disease, and cancer. On the health end, hydrogen offers the long-sought-after fountain of youth because it puts the rocket power of hydrogen directly into one's cells. Everyone gets something from hydrogen because life cannot exist without it.

Hydrogen inhalation is a medicine that doubles as a health practice. Some users already regard H2 inhalation as an essential act of life, like eating, drinking, and exercising. Meaning it is always helpful to have more hydrogen in your tank.

Hydrogen inhalation is like getting two or more intravenous Vitamin C treatments a day without toxicity for people with late-stage cancer. In the ICU, it can be administered 24/7, meaning until the patient gets up and out of bed. Some doctors already know what antioxidants like Vitamin C can do for sepsis patients. Hydrogen offers those on death door even more relief.

Hydrogen gas and hydrogen impregnated water offer doctors and patients alike therapeutic strategies that promote health and quality of life in clinical environments. Besides, hydrogen also has beauty applications for your skin. Want to look young again?

Higher levels of hydrogen protect your DNA against oxidative damage. It suppresses the single-strand breakage of DNA caused by reactive oxygen species free radicals. Hydrogen also repairs oxidative damage to RNA proteins. Anyone suffering from diabetic foot or neuropathy will be delighted with the inhalation of hydrogen gas over a few months. Like carbon dioxide, hydrogen can be pumped into a bagged arm or leg to treat syndromes like gangrene and skin cancer topically.

My writings on hydrogen are enthusiastic, and I am not alone in this. "It is not an overstatement to say that hydrogen's impact on therapeutic and preventive medicine could be enormous in the future," write medical scientists.[1]

If I have anything to do with it, hydrogen will eventually assume its rightful place next to oxygen. Hydrogen in life-threatening situations can even be miraculous Over the long haul, it just keeps working like a solar wind, filling our sails and pushing our ship of life away from the rocky shoals of sickness and death.

Hydrogen is not an instant magic wand. If one is looking for dramatic effects, one needs to spend dramatic amounts of time inhaling hydrogen gas. If your life is on the line, think about a continuous application, even breathing hydrogen all night while you sleep. If one wants to be aggressive with anti-aging treatments, think of two hours a day to eventually become young again.

Although hydrogen is not a miracle medicine, one can expect many miracles to happen. I was talking with Tyler LeBaron of the Molecular Hydrogen Institute, saying, "Too bad hydrogen therapy does not offer enlightenment, but for the very sick, I imagine it might seem that way." And he responded, "Yes, for sick people, it does offer enlightenment, and even for those that are not sick, it may improve their cognitive function."

Hydrogen is by far the cleanest form of energy and is also the safest medicine, meaning it has no harmful effects, only good ones. Going through life and facing disease is more comfortable if we fill our bodies and cells with hydrogen if we pay attention to oxygen and blood CO_2 levels simultaneously.

Hydrogen is the ultimate medicine for high performers and the chronically ill alike. It will be hard to die from the flu if one is breathing oxygen and hydrogen together. Hydrogen gas will help anyone stay alive longer, no matter what their medical situation.

Introduction to Hydrogen Medicine.

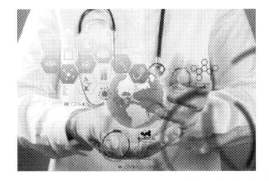

Modern medicine is discovering how brilliant, safe, and effective hydrogen can be. Molecular hydrogen brings a revolution to treatment. The science is irresistible, and mounting clinical experience points to hydrogen as the most straightforward, safest medicine in existence. After experimenting with toxic pharmaceuticals, heavy metals, and radiation over the last century, we now find hydrogen gas and hydrogen water, making headway into the mainstream of medicine where it is saving lives.

Hydrogen to the Rescue.

What is the first thing the fire department does when its trucks arrive at a fire? They pull out the hoses, connect to the hydrants, and pour tons of water onto the nastiest fires. In the medical world, the equivalent is hydrogen, which can be flooded into a dying body to put out the worst flames of inflammation and oxidative stress.

As you shall see throughout this book, molecular hydrogen (H_2) functions as an extensive protector against oxidative stress, inflammation, and allergic reactions.[2] Molecular hydrogen has proven valuable and convenient as a novel antioxidant and modifier of gene expression where oxidative stress results in cellular damage.

> A man who had Type 2 diabetes was not getting results using hydrogen inhalation. When asked how he was doing his hydrogen therapy, he said he was inhaling the gas but only drinking a little water. After being urged to drink the water more regularly, he bought three 17 oz. stainless thermos bottles drinking that much hydrogen water each day. Before, his blood sugar was out of control, and doctors had to increase insulin. But since he has been combining H_2 inhalation with drinking the water daily, his blood sugar has come down and normalized after about a week.

Hydrogen can cross the blood-brain barrier, enter the mitochondria, and "even can translocate to the nucleus under certain conditions," reports Dr. Brandon J Dixon in *Medical Gas Research*. Because of its size, molecular hydrogen can do a lot; more massive antioxidants cannot. Studies have shown that hydrogen also exerts anti-apoptotic and cytoprotective properties that are beneficial to the cell.

There are no safety issues with hydrogen; it has been used for years in gas mixtures for deep-sea diving and numerous clinical trials without adverse events. There are no warnings in the literature on its toxicity or long-term exposure effects.

Hydrogen will change the landscape of a person's present condition, and that's why doctors are taking notice. Its systemic properties and penetration abilities allow hydrogen to be useful for poor blood flow and other situations that limit many systemic treatments.

Hydrogen protects us from harm.

The first report demonstrating the benefits of drinking hydrogen water in patients receiving radiation therapy for malignant tumors affirms how helpful hydrogen is for radiation exposure. This finding provides the foundation for a clinically appropriate, effective, safe strategy for delivering hydrogen gas to mitigate radiation-induced cellular injury.[3]

Hydrogen also helps protect us from the cellular damage caused by cell phones, Wi-Fi, and all the EMF pollution that is increasingly plaguing us, which is about to worsen with the implementation of G5.

In a World of Increasing Radiation.

Chronic oxidative stress is the leading cause of post-radiation effects, including cancer.[4] Cellular exposure to ionizing radiation leads to oxidizing events that alter the atomic structure through direct radiation interactions with target macromolecules. Further, the oxidative damage may spread from the targeted to neighboring, non-targeted bystander cells. In irradiated cells, these reactive species' levels may be increased due to perturbations in oxidative metabolism and chronic inflammatory responses, thereby contributing to the long-term effects of exposure to ionizing radiation on genomic stability.[5]

We are constantly faced with oxidative stress, and reducing it is one of the most important things we can do for our health. Because background radiation is increasing, so are our levels of oxidative stress.

Cosmic radiation induces DNA and lipid damage associated with increased oxidative stress and remains a major concern in space travel. Now that we are in the beginnings of a Grand Solar Minimum, flying has become more dangerous due to these increasing cosmic rays penetrating our atmosphere. Hydrogen for astronauts can potentially yield a novel and feasible preventative/therapeutic strategy for radiation-induced events, either by inhalation or drinking hydrogen-rich water.

The Fire of Life

Just as every sun needs mountain ranges of hydrogen every minute to run themselves, humans too run on hydrogen. Supplying more, in the form of molecular hydrogen gas and water, ignites a fire in us. Medical scientists believe a healing fire is even good for impossible to cure diseases like muscular dystrophy and many other conditions that modern medicine has failed to find answers for.

This book presents a chapter dedicated to hydrogen as a fuel. The fire of life gets stimulated by hydrogen. If one needs healing, that fire will be directed precisely to the place it is most needed. We can create a perfect hydrogen flame by balancing the three gases so that the mitochondria fire up ATP production. With enough hydrogen and oxygen, the mitochondria will burn without oxidative prejudice. The hydrogen diminishes the fires of oxidative stress even as it is the fuel that lights the fire.

Hydrogen is one of the primordial elements that fueled the development of all life on Earth. Human beings cannot live without hydrogen. While science refers to us as carbon-based life forms, man is also a hydrogen-based life form. When plants absorb sunlight, they store negatively charged hydrogen ions through the process of photosynthesis. When you eat unprocessed plants, your body's cells utilize those plants' nutrients with the hydrogen ions' electrical charge. When your body burns hydrogen and oxygen, it generates the energy we need to live.

All living things must have hydrogen to sustain life. The human body must breathe to get oxygen and must eat and drink to get hydrogen. The primary sources of hydrogen ions for the human body are fresh uncooked plants, fruits, vegetables, and water. Now we can inhale directly molecular hydrogen gas. Dissolved in water, we can drink it, or doctors can inject it.

It is appropriate to discuss a little what hydrogen will not do. It will not fix a broken bone. It will not resolve a busting appendix though it should be used during and after surgery. It will not resolve the conflicts and stress people feel from personality problems. Though it can strengthen a person's sexuality, it will not touch on negative patterns of behavior.

It will not, all by itself, balance the three nervous systems of Ayurvedic Medicine Vata, Pitta, and Kapha. Hydrogen will not balance meridian energies nor affect the chakras in any meaningful way. Though after long-term usage I might be proven wrong.

As I write these words just before publication I have been sleeping all night every night on my new hydrogen/oxygen machine (Brown's Gas) and I feel like I am about to blast off. Have not looked or felt so good in ten years.

Though hydrogen is helpful in all emergencies, it will not substitute for magnesium if a person has a heart attack. Magnesium will work almost instantly in situations where hydrogen takes time.

Hydrogen gas and hydrogen water are helpmates. They are not a cure for anything though they might help everything.

Hydrogen for Every Aliment.

Treating cancer with hydrogen, oxygen, and carbon dioxide is an approach that treats the fundamental reasons cancer cells form and get aggressive in the first place. Cancer should not be treated as a genetic disease; it is more like a metabolic disease.

Hydrogen is suitable for every aliment known to humankind, just as it is essential for every star in the universe. The sicker a person is more they will experience the benefits of hydrogen. Hydrogen can be flooded into the body to put out the worst flames of inflammation and oxidative stress. In Hydrogen Medicine, we flood the body with the three primary gases—hydrogen, oxygen, and carbon dioxide—as a first course of action in all dire medical situations. The same goes for any chronic or acute condition like the flu. The longer one wants to live, the more one supplements these primary gases. The most powerful healing/medical/anti-aging device in the world is a hydrogen oxygen inhaler.

Hydrogen gas therapy is a promising novel treatment for emergency and critical care medicine. It exerts a therapeutic effect in a wide range of disease conditions: From acute illness such as ischemia-reperfusion injury, a shock to chronic diseases such as metabolic syndrome, rheumatoid arthritis, and neurodegenerative diseases.

Concerning various emergency and critical care aspects, researchers report hydrogen is helpful for acute myocardial infarction, cardiopulmonary arrest syndrome, sepsis, contrast-induced acute kidney injury, and hemorrhagic shock. Hydrogen gas has even been used to attenuate oxidative stress in a rat model of subarachnoid hemorrhage.

When you are under tremendous stress and must continue at high-performance, hydrogen medicine should be right by your bed, sitting under your desk at the office, or both.

Inhaling hydrogen gas (with oxygen included) with a hydrogen inhalation device, from a cell's point of view, is like standing under a waterfall that invigorates even as it cools. Hydrogen will help anyone stay alive longer, no matter what situation.

It is time for the world to embrace hydrogen as a clean energy carrier to save us from the pollution nightmare affecting every urban center in the world; it is also time for modern medicine to wake up to the power and purity of hydrogen as medicine. Hydrogen produces zero toxic emissions when used for energy; it has zero side effects when used as a medicine.

Hydrogen belongs in ambulances, emergency centers, and Innovative ICU departments (Intensive Care) right alongside oxygen, which does not quite work as well as needed without hydrogen being present in abundance. Imagine water without hydrogen, and you will get the significance of combining oxygen and hydrogen gas as the ultimate way to stabilize critical patients.

It is hard to lose with hydrogen. H2 inhibits cell viability, migration and invasion, and catalyzed cell apoptosis. Hydrogen helps us fight cancer directly, minimizing its spread. It is helpful as an effective adjunct to radiation and chemotherapy treatments moderating their toxic effects by cooling oxidative stress.

Hydrogen Medicine is revolutionary; it holds the potential to save lives, reduce suffering, and make one beautiful and young again through its anti-aging effects. Importantly, with the increases of human-made nuclear radiation (think Fukushima), medical tests that use radiation, substantial increases in cosmic rays (caused by Grand Solar Minimum), and the coming of G5 telecommunications, which has not been tested for safety, hydrogen gas therapy is becoming essential to mitigate the increases of oxidative stress.

A hydrogen/oxygen gas machine is probably the first piece of medical equipment that one should invest in, for it offers a fundamental treatment for almost all disease conditions. It is not a cure-all and should be used in the context of a complete protocol, with particular attention paid to increasing CO2 levels through sodium and potassium bicarbonates and slow breathing.

New Advances in Gas Therapy.

Ambulance crews have often regarded oxygen as something approaching a wonder drug. Oxygen has always been a lifesaving treatment, and now doctors and patients can combine oxygen with hydrogen to achieve even more. This pair will have a substantial impact on medicine.

It is hydrogen that allows the body to function and breathe under stress! The United States Navy developed procedures allowing dives between 500 and 700 meters (1650 to 2300 feet) in depth while breathing gas mixtures based on hydrogen, called hydroxy (hydrogen-oxygen) or hydreliox (hydrogen-helium-oxygen).

Although the first reported use of hydrogen seems to be by Antoine Lavoisier (1743–1794), who had guinea pigs breathe it, the first uses of this gas in deep-sea diving are attributed to the Swedish engineer Arne Zetterström in 1945. Breathing a mixture of 96% hydrogen and 4% oxygen, he could dive deeper than anyone before.

Hydreliox is a breathing gas mixture of helium, oxygen, and hydrogen. For the Hydra VIII mission at 50 atmospheres of ambient pressure, the combination was 49% hydrogen, 50.2% helium, and 0.8% oxygen.

It is hard to imagine a more stressful situation than being that deep underwater. If hydrogen can keep us alive under great stress at bone-crushing depth, think of what it can do if one dies or suffers from a chronic disease.

Medical Gases.

Heliox is a breathing gas composed of helium (He) and oxygen (O2). Heliox is a medical treatment for patients with difficulty breathing. The mixture generates less resistance than atmospheric air when passing through the lungs' airways, requiring less breathing effort. Heliox has been used medically since the 1930s. Although the medical community initially adopted it to alleviate upper airway obstruction symptoms, its range of medical uses has since expanded because of the low density of the gas.

Through the Bohr Effect,[6] medical scientists have known the importance of carbon dioxide (CO2). However, hydrogen allows for quicker healing and recovery than O2 and CO2.

Medical gases trigger naturally occurring physiological responses, enhancing the human body's preventive and self-healing capabilities. Medical gases include carbon dioxide, oxygen, nitrogen, nitric oxide, helium, and, most recently, hydrogen. Medical gases are used on their own or in combination as therapeutic effects or insufflation during surgery.

Inhaled Nitric oxide works by relaxing smooth muscles to dilate blood vessels, especially in the lungs. Nitric oxide, together with a mechanical ventilator, treats respiratory failure in premature infants. Increased levels of CO2 also dilate blood vessels while positively affecting the oxygen disassociation curve. Molecular hydrogen (H2) is an inert and non-functional gas in our body, but this is not correct. H2 reacts with strong oxidants such as hydroxyl radicals in cells.

Pediatric and neonatal patients have an assortment of physiologic conditions that may require adjunctive inhaled gases to treat a wide variety of diseases. Inhaled nitric oxide, helium-oxygen mixtures, inhaled anesthetics, hypercarbia mixtures, hypoxic mixtures, and hydrogen are used to alter physiology to improve patient outcomes.[7]

Xenon is another medical gas capable of establishing neuroprotective, inducing anesthesia, and serving in modern laser technology and nuclear medicine as a contrast agent. Despite its high cost, lack of side effects, safe cardiovascular and organ protective profile, and neuroprotective role after hypoxic-ischemic injury (HI), doctors favor its applications in clinics.

Hydrogen is Serious Medicine for Cancer.

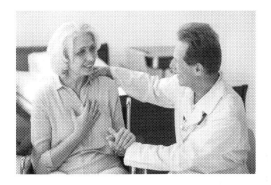

The most recent research in China suggests that molecular hydrogen may exert anti-tumor roles in ovarian cancer by suppressing the proliferation of Cancerous stem-like cells and angiogenesis. Six weeks of hydrogen inhalation in animal studies significantly inhibited tumor growth, decreasing mean tumor volume (32.30%). Hydrogen treatment reduced the expression of CD34[8] (74.00%), demonstrating its anti-angiogenesis effects.[9] The *in vitro* study showed that hydrogen treatment significantly inhibits cancer cell proliferation, invasion, migration, and colony formation.

In 1975, an impressive study demonstrated that hyperbaric molecular hydrogen therapy could be a possible cancer treatment. In this study, the researchers showed that exposing mice with skin cancer (tumors) to 2.5 percent oxygen (O2) and 97.5 hydrogens (H2) for two weeks produced a significant regression of the mice tumors. "After exposing mice10-days with hydrogen-oxygen therapy, tumors turned black. Some dropped off; some seemed shrunk at their base and ready to be pinched off. The mice appeared to suffer no harmful consequences."[10]

A similar effect of hyperbaric hydrogen on leukemia was reported in 1978.[11] An anti-inflammatory effect of hyperbaric hydrogen on a mouse model with chronic liver inflammation was also published in 2001.[12]

Hydrogen is a new method for treating cancer.

Survey of hydrogen-controlled cancer: a follow-up report of 82 advanced cancer patients with stage III and IV cancer treated with hydrogen inhalation with the following results:

"After 4 weeks of hydrogen inhalation, patients reported significant improvements in fatigue, insomnia, anorexia and pain.

Furthermore, 41.5% of patients had improved physical status, with the best effect achieved in lung cancer patients and the poorest in patients with pancreatic and gynecologic cancers.

The greatest marker decrease was in achieved lung cancer and the lowest in pancreatic and hepatic malignancies.

Of the 80 cases with tumors visible in imaging, the total disease control rate was 57.5%, with complete and partial remission appearing at 21-80 days (median 55 days) after hydrogen inhalation.

The disease control rate was significantly higher in stage III patients than in stage IV patients (83.0% and 47.7%, respectively), with the lowest disease control rate in pancreatic cancer patients."

Conclusion.

"In patients with advanced cancer, inhaled hydrogen can improve patients' quality-of-life and control cancer progression."[13]

Oxidative stress in the cell results from excess reactive oxygen species (ROS). Acute oxidative stress may result from various conditions, such as vigorous exercise, inflammation, ischemia, reperfusion (I/R) injury, surgical bleeding, and tissue transplantation. Chronic/persistent oxidative stress is closely related to the pathogenesis of many lifestyle-related diseases, aging, and cancer.

The story of hydrogen as medicine began in 2007 when Ohsawa and colleagues discovered H2 has antioxidant properties that protect the brain against I/R injury and stroke by selectively neutralizing hydroxyl radicals.[14] If H2 reduces oxidative stress, we can also say H2 protects DNA because oxidative stress damages cellular DNA, leading to mutations. ROS causes oxidative DNA and protein damage as well as damage to tumor suppressor genes. Unlike other antioxidants, the tiny size of H2 molecules allows them to penetrate bio-membranes and diffuse into the mitochondria and nuclei, thereby protecting the nuclear DNA and mitochondria.

ROS causes damages to cell structures via oxidative stress. While a cancer drug like Cisplatin effectively kills cancer cells, it also harms other cells in your body through oxidative stress. How do you prevent the killing of good cells in your body while the cancer drug is doing its job to kill the cancer cells? The answer is hydrogen gas; introduced continuous antioxidant protection reducing the oxidative stress to the non-cancerous cells in your body. And as we shall see later in this book, sodium, potassium, and magnesium bicarbonates should be used in every cancer case, giving us instant command of CO2 levels in the blood and increasing oxygen delivery to the cells.

Human tumor cells produce more reactive oxygen species (ROS) than non-transformed cell lines. Elevated oxidative stress increases cell proliferation, DNA synthesis, cellular migration, invasion, tumor metastasis, and angiogenesis.[15] Anything that will reduce ROS consistently is going to be ideal for cancer treatment.

Hydrogen suppresses the VEGF (Vascular Endothelial Growth Factor), a key mediator of tumor angiogenesis (the development of new blood vessels), by the reduction of excessive ROS (oxidative stress) and through the downregulation of ERK (key growth factor needed for cellular division).[16]

More Science on Hydrogen and Cancer

"Molecular hydrogen reduces the growth of human tongue carcinoma cells HSC-4 and human fibrosarcoma cells HT-1080 but did not compromise the development of regular human tongue epithelial-like cells.

H2 has been shown to reduce the exaggerated expressions of MMP genes (MMP proteins are involved in multiple functions in cells, including cell proliferation, cartilage synthesis, apoptosis, angiogenesis, etc.). We know that cancerous cells have a higher expression of MMP genes leading to tumor invasion and tumor angiogenesis.[17],[18]

"ERW [hydrogen water] causes telomere shortening in cancer cells and suppresses tumor angiogenesis by scavenging intracellular ROS and suppressing the gene expression and secretion of vascular endothelial growth factor. Also, ERW induces apoptosis together with glutathione in human leukemia HL60 cells.[19] "Treatment with H2 water increased the expression of p-AMPK, AIF, and Caspase 3 (cell apoptosis pathways) in colon 26 cells. Thus, H2 water resulted in cell apoptosis mediated by the AMPK pathway in colon 26 cells".[20]

Molecular hydrogen may protect and retard the development of thymic lymphoma in mice. "The radiation-induced thymic lymphoma rate in an H2 group was significantly lower than in the control group. H2 treatment significantly increased the latency of lymphoma development after the split-dose irradiation. These data indicated that H2 protects mice from radiation-induced thymic lymphoma in BALB/c mice."[21]

Hydrogen for Cancer Patients Receiving Radiotherapy and Chemo

Molecular hydrogen protects healthy tissue/cells from anticancer drugs and against chemotherapy drugs. Hydrogen has the potential for improving the quality of life during chemotherapy by efficiently mitigating the side effects of Cisplatin."[22]Hydrogen alleviates nephrotoxicity-induced Cisplatin without compromising anti-tumor activity in mice.

Nearly half of newly diagnosed cancer patients receive radiotherapy. While radiotherapy destroys malignant cells, it also affects normal surrounding cells. Acute radiotherapy side effects include fatigue, nausea, diarrhea, dry mouth, loss of appetite, hair loss, sore skin, and depression. The side effects are associated with increased oxidative stress and inflammation due to ROS generation during radiotherapy. The consumption of hydrogen-rich water for six weeks reduced reactive oxygen metabolites in the blood and maintained blood oxidation potential. During radiotherapy, quality of life was significantly improved in patients treated with hydrogen-rich water than patients receiving placebo water.[23]

Hydrogen Inhalation Treats Brain Cancer.

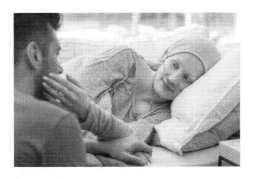

An important case study of a 47-year-old woman suffering from multiple brain tumors showed how critical hydrogen medicine would be in the future of medicine and modern oncology. In another study of a 44-year-old woman diagnosed with lung cancer with multiple metastases in November 2015, targeted oral drugs were initiated after removing brain metastases. Most lesions remained stable for 28 months. In March 2018, multiple intracranial metastases, hydrocephalus accumulation in the third ventricle and lateral ventricles, and metastases in bone, adrenal gland, and liver were noted. Hydrogen-gas monotherapy started to control the tumor a month later. After four months, the size of multiple brain tumors and hydrocephalus in the third ventricle and lateral ventricles were reduced significantly. After one year, all brain tumors had disappeared, though there were no significant changes in metastases in the liver and lung. After standard treatments had failed, these data show that hydrogen-gas monotherapy elicited effective control of tumors (especially those in the brain) and extended survival time.

Non-small-cell lung cancer (NSCLC) is the second most frequently diagnosed tumor worldwide and is the leading cause of cancer-related deaths. Under orthodox oncology, 5-year survival is 3.6%. There are no significant changes in 5-year survival for patients with NSCLC metastases to the brain in the last three decades.

The gas generated consists of 67% hydrogen and 33% oxygen. Using a special mask, the patient continues to inhale hydrogen for 3–6 hrs a day at rest, with no interruption even after the apparent relief of symptoms. After four months, most brain metastases disappeared.

"Overall, H2 reduces the risk of lifestyle-related oxidative stress by reacting with strong reactive oxygen/nitrogen species in cell-free reactions. It is easy to apply H2 in cases of oxidative stress, inflammation, and tumors. Due to the lack of adverse effects and the high efficacy for most pathogenic statuses involved, H2 gas and H2 water are increasingly being accepted as promising candidates for therapeutic approaches. We hypothesize that H2 gas inhalation and oral administration of H2 water could protect against inflammation in oxidative stress-related cancer, and thus improve the anti-tumor effect in the clinical management of cancer."

"We present the case of a 72-year-old female patient with gallbladder cancer (GBC) who developed in situ recurrence and liver metastases nine months after irreversible electroporation ablation and oral chemotherapy, which failed to control the progression of the disease. The patient further developed metastases in the lymph nodes around the head of the pancreas. The patient had severe anemia, requiring weekly blood transfusions. The gallbladder tumor invaded the descending part of the duodenum, causing intestinal leakage and hepatic colonic adhesion. Case summary: The patient refused other treatments and began daily hydrogen inhalation therapy. After one month of treatment, the gallbladder and liver tumors continued to progress, and intestinal obstruction occurred. The intestinal obstruction was gradually relieved after continuous hydrogen therapy and symptomatic treatments, including gastrointestinal decompression and intravenous nutrition support. Three months after hydrogen therapy, the abdominal cavity's metastases gradually reduced in size. Her anemia and hypoalbuminemia were corrected, lymphocyte and tumor marker levels returned to normal, and the patient could resume a normal life. Conclusion: This is the first report of an efficacy and safety study about hydrogen therapy in a patient with metastatic GBC and a general critical condition, who has remained stable for more than four months."

Treating Lung Cancer With Hydrogen, Bicarbonates, Glutathione, and Iodine.

No matter how they spin the statistics, over 50% of lung cancer patients die from their disease or from the treatment they receive from oncologists. The American Cancer Society's estimates for lung cancer in the United States for 2021 are:

- About 235,760 new cases of lung cancer (119,100 in men and 116,660 in women)
- About 131,880 deaths from lung cancer (69,410 in men and 62,470 in women)

The National Foundation For Cancer Research tells us that only 16% of lung cancers are identified early, making treatment and survival much more difficult. Each year, hundreds of thousands of Americans are diagnosed with lung cancer. Tragically, lung cancer is amongst the deadliest form of cancer, claiming more lives each year than breast, prostate, and colon cancers combined.

Because most lung cancer cases are diagnosed in the metastatic stage, chemotherapy is the standard treatment option. Unfortunately, chemotherapy is not very effective. Some lung cancer cells are intrinsically resistant to chemotherapy, while others develop resistance and then multiply.

What the National Foundation for Cancer Research says is, "The lack of other treatment options greatly contributes to lung cancer's shocking death toll." This is not true or stated incorrectly. It is the narrow-minded stubbornness of oncologists who refuse to look at other options that exist that causes lung cancer's shocking death toll. In modern medicine in general and especially for oncologists, if they do not know about alternative treatments, it means they do not exist—naked hubris or, in terms of cancer, deadly arrogance.

The best way of treating lung cancer is to treat the lungs themselves, though because most lung cancer patients are suffering from metastasis, systemic treatments are also required. Direct treatments to the lungs can be considered topical/transdermal treatments because the lungs are the inner skin that separates the body from the environment.

The main focus of this presentation will be split between hydrogen and oxygen gas, which can be administered many hours a day, or continuously in the later most deadly stages, and the nebulization (getting medicines directly into lung tissues) using sodium bicarbonate, glutathione, iodine, and magnesium chloride. Though it is too early to provide statistics, such a treatment combination is backed by logical medical science. Therefore, one cannot go wrong with such treatments.

The most revolutionary treatment for lung cancer is hydrogen gas. Hydrogen therapy can control tumor progression and alleviate the adverse events of medications in patients with advanced non-small-cell lung cancer, medical scientists in China have concluded.

Dr. Ji-Bing Chen et al. of Fuda Cancer Hospital of Jinan University administered H_2 inhalation for 4–5 hours per day. "During the first five months of treatment, the control group's prevalence increased gradually, whereas that of the four treatment groups decreased gradually. After 16 months of follow-up, progression-free survival of the control group was lower than that of the H_2-only group and significantly lower than that of H_2 + chemotherapy, H_2 + targeted therapy, and H_2 + immunotherapy groups. In the combined therapy groups, most drug-associated adverse events decreased gradually or even disappeared. H_2 inhalation can be used to control tumor progression and alleviate the adverse events of medications for patients with advanced non-small-cell lung cancer."[24]

Dr. Jinghong Meng et al. found that hydrogen gas represses the progression of lung cancer. Their study concluded that H2 inhibits lung cancer progression via down-regulating CD47, which might be a powerful method for lung cancer treatment.[25]

Dr. Dongchang Wang_et al. showed that H_2 inhibited cancer cell viability, migration, and invasion and catalyzed cell apoptosis. "All data suggested that H_2 inhibited lung cancer progression through down-regulating SMC3, a regulator for chromosome condensation, which provided a new method for the treatment of lung cancer."[26]

"In addition, inhibition of the proliferation, migration, invasion, and promotion of the apoptosis of cancer cells was found when hydrogen gas was administered. The experimental animal assay demonstrated that the tumor weight in the H_2 group was significantly smaller than that in the control group."

Dr. Sai Li et al. found that hydrogen possesses multiple bioactivities, including anti-inflammation, anti-reactive oxygen species, and anti-cancer. In addition, growing evidence has shown that hydrogen gas can alleviate the side effects caused by conventional chemotherapeutics or suppress the growth of cancer cells and xenograft tumors, suggesting its broad potent application in clinical therapy.[27]

As the lightest molecule in nature, hydrogen gas exhibits appealing penetration property, as it can rapidly diffuse through cell membranes. Although hydrogen gas was studied as a therapy in a skin squamous carcinoma mouse model back in 1975,[28] its potential in medical application was not explored until 2007. *Hydrogen Medicine* is new, but that is no excuse for it not being recognized as the ideal and most essential foundational treatment for lung cancer.

Back in 1975, mice with squamous cell carcinoma were exposed for periods up to 2 weeks to hydrogen to see if this free radical decay catalyzer would cause regression of the skin tumors. "After a first 10-day period of exposure of the mice to the hydrogen-oxygen therapy, it was found qualitatively that the tumors had turned black, that some had dropped off, that some seemed to be shrunk at their base and to be in the process of being "pinched off," and that the mice appeared to suffer no deleterious consequences." However, this effect was not observed when the mice were exposed to other gases such as helium and oxygen. At the end of their article, the authors predicted that this anti-cancer activity might be due to the reaction of hydrogen with the most potent oxidant known to humankind, i.e., the OH radical.

It's hard to talk about what to anticipate at the end of life with lung cancer. Yet many people wish for some idea of what to expect at the final stage of the journey for their loved ones or themselves. How any person experiences the end of life will be different if hydrogen is administered along with the other natural substances and therapies listed below. Hydrogen inhalation therapy is ideal for all stages of lung cancer, not only because of its efficacy but because it is easily administered.

Oxygen pulls the rug out from under cancer cells and tumors by removing the fundamental condition that makes them virulent. Bicarbonates do the same thing so, using oxygen and bicarbonate together is lethal to cancer cells.

In addition, for the most severe cases, hydrogen and oxygen gases (Brown's Gas) can be administered around the clock. Hydrogen Inhalation machines have proven themselves in ICU departments for a variety of life-threatening situations. Hydrogen, for the dying, is like a wind that blows a person away from death's door. Hydrogen can cheat the angel of death, so for lung cancer patients, it is crucial.

The scientific literature and the rapidly emerging science indicate a great potential of Hydrogen gas inhalation in oncology. Hydrogen Medicine has already shown itself to be effective for brain cancer—which is especially important for lung cancer patients and breast cancer patients for metastasis often goes to the brain.

Nebulization – Transdermal Treatments Into the Lungs

In some countries, nebulizers are given to people by prescription only because they provide direct access to the bloodstream. This is an indication that this is serious medicine we are dealing with, so caution is advised. With nebulizers, we, in part, get the same effect as with injections; medications quickly diffuse directly into the bloodstream. Thus a nebulizer can be a lifesaver.

People with a lung ailment do better when taking drugs by nebulization instead of orally. This is because the embattled system doesn't need to break down the medications in the stomach and then deliver them to the lungs through the bloodstream. Instead, nebulization medicines get sprayed directly onto the lung tissues to be absorbed locally by the lung and brachial cells.

Dr. Shallenberger says, "A nebulizer converts a liquid into tiny bubbles that are so tiny that they can only be seen under a microscope. When these bubbles come out of the nebulizer, they are so small that they look just like smoke. And that's the magic of a nebulizer. The bubbles are so small that they can be inhaled deep down into the deepest regions of the lungs without any discomfort or irritation. It's a great way for asthmatics to get the medication they need to open up their lungs."

Glutathione

Dr. Michelle Alpert says, "Because oral glutathione is not well absorbed, I have also begun to experiment with nebulized glutathione, which patients can take at home between detox drips. According to a study in Alternative Medicine Review in 2000, nebulized glutathione has had remarkable success in emphysema and other lung disorders such as asthma and bronchitis. It appears that inhalation may have a systemic effect. Some patients are having even greater success with this combination."

I have been nebulizing glutathione with baking soda for my chronic bronchitis. I was astounded the first time I used it. I did not have any problems coughing up mucus for 24 hours, and I had a lot less discharge over the next few days.

N. Goodfellow

Glutathione, the most critical antioxidant in the body, is the place where sulfur and selenium meet up to protect us from cancer. The immune system cannot function properly without it, and antioxidants such as vitamins C and E rely on it to work appropriately within the body. The glutathione cancer connection is well established. Patients with cancer, chronic severe illness, AIDS, and over 60 other diseases have reduced glutathione levels. In addition, glutathione plays a specific role in detoxifying many well-known cancer-causing and cell-damaging substances in our environment.

Over 98,000 scientific studies and articles on glutathione are recorded in PubMed, the official U.S. Government library of medical research. Those articles reveal the remarkable role glutathione plays in every cell's protection and function in the human body and optimal health and function support. However, they also show the terrible consequences of low glutathione levels and how those lower levels accelerate the aging process.

Magnesium and Glutathione

Pure magnesium oil can be used in a nebulizer. Glutathione requires magnesium for its synthesis. Glutathione synthetase requires a-glutamyl cysteine, glycine, ATP, and magnesium ions. Data demonstrates glutathione's direct action in vivo and in vitro to enhance intracellular magnesium and a clinical linkage between cellular magnesium, GSH/GSSG ratios, and tissue glucose metabolism.

According to Dr. Russell Blaylock, low magnesium is associated with dramatic increases in free radical generation and glutathione depletion, which is vital since glutathione is one of the few antioxidant molecules known to neutralize mercury.

I recommend L-Glutathione Plus for nebulization because it is specifically engineered for enhanced absorption using pharmaceutical grade (OPITAC) reduced L-glutathione and sodium bicarbonate. When mixed with distilled or DI water or saline solution, Reduced L-Glutathione Plus™ becomes isotonic, making it comfortably ideal for lung, nasal, and other soft tissue contacts. No fillers, excipients, preservatives, or additives.

Bicarbonate

Now we have more reason than ever to focus on bicarbonate therapy for cancer. A new Ludwig Cancer Research study said, "If you want to clean cancer's clock—that is, defeat it decisively—you may want to really clean it—that is, restore it to proper working order. In sympathy with the tick-tock of their internal circadian mechanisms, only cancer cells that remain active in sympathy with the tick-tock of their internal circadian mechanisms remain susceptible to cancer therapies. So, how might these circadian mechanisms be kept in motion? Sodium bicarbonate now promises to awaken cancer cells that have gone to sleep deep inside tumors, where oxygen deprivation and acidic conditions go hand in hand. By buffering against acidification, sodium bicarbonate rescues circadian oscillation."

Virtually every cancer patient will benefit from sodium bicarbonate because it treats the low oxygen acid conditions universally found in cancer patients. Sodium bicarbonate shrinks tumors. (Bicarbonate inhibits spontaneous metastases (Robey 2009). 'Bicarbonate Increases Tumor pH and Inhibits Spontaneous Metastases' according to medical scientists. NaHCO3 therapy significantly reduced the formation of hepatic metastases following intrasplenic injection, suggesting that it did inhibit extravasation and colonization.

"Studies conducted at the University of Bari in Italy demonstrated that a hallmark of all tumors, regardless of their origin or background, is their acidic environment. Tumor progression increased with an acidic pH and hypoxia, or a low oxygen level," writes Dr. Veronique Desaulniers. Thus, the perturbation in pH dynamics rises very early in carcinogenesis and is one of the most common pathophysiological hallmarks of tumors."

"The results of a study suggest that tumor cells do, indeed, perform niche engineering by creating an acidic environment that is non-toxic to the malignant cells but, through its negative effects on normal cells and tissue, promotes local invasion."

Tumor invasion did not occur in regions with normal or near-normal pH. Immunohistochemical analyses revealed cells in the invasive edges expressed the glucose transporter GLUT-1 and the sodium-hydrogen exchanger NHE-1, both associated with peritumoral acidosis. In support of our findings, oral administration of sodium bicarbonate was sufficient to increase peritumoral pH and inhibit tumor growth and local invasion in a preclinical model, supporting the acid-mediated invasion hypothesis.

Increased systemic concentrations of pH buffers lead to reduced intratumoral and peritumoral acidosis. Oral NaHCO3 selectively increased the pH of tumors and reduced spontaneous metastases in mouse models of metastatic breast cancer. NaHCO3 therapy also reduced the rate of lymph node involvement and significantly reduced the formation of hepatic metastases. In addition, acid pH increased the release of active cathepsin B, an essential matrix remodeling protease.

The more alkaline you are, the more oxygen your fluids can hold and keep. Oxygen also buffers/oxidizes metabolic waste acids helping to keep you more alkaline. The quickest way to increase oxygen and pH is through sodium bicarbonate administration, which is why bicarbonate has always been a mainstay in emergency rooms and intensive care medicine. But, of course, when we increase oxygen and pH levels, we simultaneously increase cellular voltage.

"The Secret of Life is both to feed and nourish the cells and let them flush their waste and toxins," according to Dr. Alexis Carrell, Nobel Prize recipient in 1912. Dr. Otto Warburg, also a Nobel Prize recipient, in 1931 & 1944, said, "If we change our internal environment from an acidic oxygen-deprived environment to an alkaline environment full of oxygen, viruses, bacteria, and fungus cannot live."

Published in the journal Cell, the above Ludwig study details how in response to acidity, cells turn off a critical molecular switch known as mTORC1 that, in ordinary conditions, gauges the availability of nutrients before giving cells the green light to grow and divide. That shuts down the cell's production of proteins, disrupting their metabolic activity and circadian clocks, pushing them into a quiescent state. "But if you add baking soda to the drinking water given to those mice, the entire tumor lights up with mTOR activity. So the prediction would be that by reawakening these cells, you could make the tumor far more sensitive to therapy."

It Makes Perfect Sense to Nebulize Iodine

In concert with its antioxidant and anti-inflammatory actions, iodine affects several molecular pathways that are part of differentiation and apoptosis in cells. Ongoing epidemiological evidence points to iodine's role in the prevention and treatment of cancers through these effects.

High rates of goiter (iodine deficiency) correlate with higher rates of cancer mortality. It has been known for over a hundred years, especially for breast and stomach cancer. Other cancers associated with low iodine goiter conditions include prostate cancer, endometrial, ovarian, colorectal, and thyroid cancer.

Cancer starts with iodine deficiencies, just as it does with low oxygenation of tissues. Dr. Brownstein lays out what we would expect to find in iodine-deficient individuals.[29] When iodine is deficient, nodules form in essential organs leading to pre-cancerous conditions and eventually leading to full-blown cancer. Brownstein says, "Iodine's main job is to maintain a normal architecture of those tissues. With iodine deficiency, the first thing that happens is you get cystic formation in the breasts, the ovaries, uterus, thyroid, prostate, and, let's throw in the pancreas in here as well, which is also increasing at epidemic rates – pancreatic cancer. Cysts start to form when iodine deficiency is there. If it goes on longer, they become nodular and hard. If it goes on longer, they become hyperplastic tissue, which is the precursor to cancer. I say that's the iodine deficiency continuum."

Brownstein continues, "The good thing about iodine is that iodine has apoptotic properties, meaning it can stop a cancer cell from just continually dividing, dividing, dividing until it kills somebody. So iodine can stop this continuum wherever it catches it and hopefully reverse it, but at least put the brakes on what is happening.

Conclusion

Almost without exception, oncologists march in lockstep with the pharmaceutical paradigm of using the most dangerous toxic treatments. Thus it pays to know more than your oncologist. I give my readers precisely that in my Conquering Cancer Course.

Whether you or someone you love has cancer, knowing what to expect can help you cope and live through the experience. From in-depth information about cancer and its causes to in-depth information about treatments, emotional aspects, tests, and a whole lot more, I offer a 100 lesson course on cancer at eighty percent off the regular price of 500 dollars. So your **cost will be only 99 dollars**. The system is part of a doctoral program at Da Vinci University and, when taken for credit, costs 1,000 Euros.

https://drsircus.com/conquering-cancer-course/

General Instructions

Procedure: The primary aim of a nebulizer is to facilitate faster and more effective absorption of the medicine by breaking down the liquid medication into very fine particles inhaled by the patient. The first step is to add the liquid medicine to the cup attached to the device. It is essential to understand that these devices accept only liquid medicine just added at the time of usage and not before that. If the doctor has prescribed more than one medicine for nebulization, make sure if they can be mixed or whether they should be taken separately. Once the medicine is put in the cup, close the cup and connect its tube to the air compressor. Turn the compressor on, and when the compressed air reaches the nebulizer cup, it will vaporize the medicine, creating a mist. The patient inhales the mist through the mouthpiece or face mask.

Take deep breaths and inhale the vapor completely. Tap the cup regularly to ensure the correct dispensation of medicine, and don't remove the mask until the mixture is used up completely. Turn on the air pump, and a mist will come from the mouthpiece. Place the mouthpiece in your mouth and breathe in slowly. At full inhalation, hold your breath for a 2-4 count to allow absorption in the lungs. If you are treating colds or sinus problems, you can also alternate breathing through your nose. It will take about 10 to 20 minutes to finish nebulization.

Hydrogen, the Fuel of Life.

Hydrogen is the most abundant element in the universe, so there's no possibility of human consumers depleting the supply. Hydrogen will save the world in terms of energy and hopefully save the medical world and many patients. According to astrophysicist David Palmer, about 75 percent of all the known matter is composed of hydrogen.

The nucleus of a hydrogen atom is a single positively charged proton—one electron orbits around. Neutrons, which are in all other elements, do not exist in the most common hydrogen form. The sun generates its energy by nuclear fusion of hydrogen nuclei into helium. In its core, the sun fuses 620 million metric tons of hydrogen each second. Of course, it is not that simple.[30] However, hydrogen gives life to our sun and all suns, spreading light all over the universe.

Hydrogen is at the Root of Creation.

There is no darkness in space; it is full of light, but we cannot see it—it looks black. If we could see, it would burn our eyes out. And we are that light. We absorb light and re-radiate like stars, and hydrogen is always at the heart of it all.

The first scientist to talk about hydrogen was Dr. Szent Györgyi, Nobel Prize laureate for discovering Vitamin C and for his work identifying the reactions that liberate energy from hydrogen. He explained one of the basic biology principles: hydrogen and oxygen interact in a delicate balance releasing energy to cells in tiny portions. Hydrogen is the only fuel the body knows. The foodstuff, carbohydrate, is only a packet of hydrogen, a hydrogen supplier, and a hydrogen donor. The combustion of hydrogen is the natural energy-supplying reaction."

Everyone knows we need oxygen to live, but oxygen's counterpart (hydrogen) is the natural fuel that burns when there is plenty of oxygen around. Today we finally see hydrogen as a clean fuel for cars.

We can see robot bees dive in and out of the water using tiny combustible rockets run on hydrogen and oxygen taken of the water they dive.[31] Gas fills a chamber in the RoboBee's interior, then lit by an internal spark, and woosh, it shoots out of the water. What is brilliant about this is the use of water as fuel. A pair of tiny electrolytic plates convert the liquid into hydrogen, igniting a rocket-like thrust.

Hydrogen is the best and most efficient energy carrier ever. It has the most amount of energy by weight out of every other combustible gas. Every fuel, be it diesel, gasoline, natural gas, propane, butane, etc., are all carbon chains of a certain length linked together by hydrogen atoms. The size of the chain determines the name and fuel type. H2, in conjunction with gasoline in your vehicle, acts as a catalyst optimizing combustion and cleaner emissions, reduced fuel costs, and reduced oil maintenance.

"The oxidation of hydrogen in stages seems to be one of the basic principles of biological oxidation. The cell would not harness and transfer the large amounts of energy released by direct oxidation to other processes. The cell needs small changes to pay for its functions without losing too much in the process. So, it oxidizes the H-atom in stages, converting the large banknote into small change," writes Szent Györgyi.

Györgyi was the first to show that the human body stores hydrogen in many of its organs. He called this 'hydrogen pooling' and identified the liver as the organ that pools the most hydrogen because it requires hydrogen to neutralize free radicals produced during detoxification. This is what hydrogen does best—neutralize free radicals and combine with them to turn them into water.

Food is a primary source of hydrogen. If it is fresh and uncooked, it provides an abundance of hydrogen. The hydrogen in food is tied up in complex molecules that need to be metabolized (broken down) to release the hydrogen. The air we breathe also contains a small amount of hydrogen, which is immediately absorbed into cells and tissues when it enters the respiratory tract. However, the amount found in the atmosphere is less than 1%.

Researchers have found a way to produce hydrogen energy using purple bacteria and electrical currents to capitalize on the organic materials we flush down the toilet every day. Purple phototropic bacteria belong to the most extensive and most diverse group of bacteria. These bacteria are photosynthetic, but they use infrared light as the energy source for their metabolism, unlike plants and algae. That gives them a color from brown to red—including purple.

The main feature of purple bacteria is their versatile metabolism. They can perform a range of metabolic reactions, including produce hydrogen gas. Purple bacteria have the problem of excess electrons from their metabolism. One way of releasing this excess is through carbon dioxide fixation like plants do. The other one is the release of electrons as hydrogen.

UK scientists have found a way to transform plastic into hydrogen. Moritz Kuehnel and colleagues at Swansea University have come up with an ingenious sunlight-driven process that uses plastic waste to produce energy in the form of hydrogen gas. A photo-catalyst, cadmium sulfide quantum dots, is added to the plastic and immersed into an alkaline solution. Sunlight then reduces the solution's water to hydrogen. Meanwhile, the plastic polymers oxidizing to small organic molecules. You can see the hydrogen gas bubbles coming right off the surface as the process works its magic.

The energy of the Future

Hydrogen is the energy of the future, and many companies are betting heavily on that future. The world's largest, most advanced hydrogen station is now in Shanghai. And now, we have a prototype of a non-polluting flying car (more like a drone) running on hydrogen that will carry five people 400 miles.

Elon Musk thinks hydrogen is silly, and that should be the end of the story, but I do not believe any of the stars in our vast universe feel the same. But when a billionaire says something, it must be true. Just look at all the rubbish that comes out of Bill Gates's mouth. People foam at the mouth over his gospel of truth on CO_2, climate change, and vaccines in and out of government. He misses the fact that all life on Earth depends on CO_2, and our cells and our atmosphere would have no oxygen without it.

Musk can build his batteries and electric cars to the end of time, but hydrogen is the fuel of existence and will be for eternity. Many believe that hydrogen plays industrial processes and power heavy-duty vehicles, including ocean-going ships and aircraft.

The Toyota car company fully invests in hydrogen. Countries like China are investing heavily in a hydrogen future. Rep. Greg Pence, R-Indiana, has recently introduced legislation to spur innovation in hydrogen energy projects in the United States. Pence introduced the measure, called the "Clean Energy Hydrogen Innovation Act," in the House.

Hydrogen is already a success. Ballard PEM fuel cell technology with the 8th generation power module launched in 2019 – has provided zero-emission power for vehicle propulsion of more than 3,400 fuel cell electric buses and commercial trucks. The Company's proton exchange membrane (PEM) has powered a cumulative total of more than 75 million kilometers on roads worldwide. That is enough to circle the globe 1870 times. Musk does not seem to care about the zero-emissions aspect of hydrogen energy.

Where does Musk think his cars are going to get charged with windmill power? In Texas, during their recent electrical crisis, it cost 900 dollars to recharge an electric vehicle. Green hydrogen is an essential part of the global transition to a more affordable, sustainable form of energy and must remain a key pillar in the energy mix. Musk should know that electric vehicles cannot fight climate change the way hydrogen cars can.

The Miracle of Molecular Hydrogen.

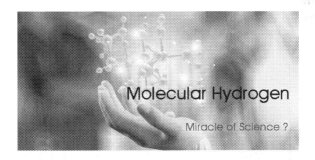

Seduced by simplicity, physicists have always found themselves fascinated by hydrogen, the simplest of atoms, which can combine to form the simplest molecules. Hydrogen has shocked, it has surprised, it has embarrassed, it has humbled - and again and again, it has guided physicists to the edge of new vistas of thought. Now it is time for hydrogen to do its magic in the field of medicine.[32]

Molecular hydrogen has anti-oxidative and anti-inflammatory properties and neuroprotective effects.[33] We see an increase in the anti-oxidative enzyme superoxide dismutase (SOD) level with hydrogen intake.[34] Hydrogen-rich saline prevents Aβ-induced neuroinflammation and oxidative stress, which may improve memory dysfunction in animal studies.[35]

Looking just at the effects of Molecular Hydrogen, we see:

- Neuro-protective
- Improves mood disorders.
- Reduces muscle fatigue, motor deficits, and muscle degeneration.
- Prevents metabolic syndrome, decreasing levels of glucose, insulin, and triglycerides: can treat diabetes.
- Antioxidant: prevents brain damage.
- Anti-inflammatory
- Protects organs.
- Lower's cholesterol and blood sugar
- Assists in weight loss.
- Enhances the mitochondrial function.
- Prevent cancer by reducing oxidative stress and suppressing tumor colony growth.
- Minimize the side effects of cancer treatments.
- Boosts skin health.
- Enhances wound healing.
- Limit's damage of transplant organs.
- Improves bladder dysfunctions.
- Is cardio protective.
- Protects and rebuilds eyes and vision.
- Prevents hearing loss.
- Combat's allergies
- Ameliorates kidney disease.
- Protects the liver.
- Promotes gut health.
- Protects the lungs.
- Protects from radiation-induced damage.
- Relieves pain.
- Is antibacterial and promotes oral health.
- Is non-toxic at high concentrations.

Millions of individuals are not getting enough hydrogen (and are suffering from it) because of mineral deficient soil, pesticides, chemical fertilizers, over-processing of foods, the addition of chemical preservatives, and drinking over-chlorinated and over-fluoridated water.

When certain chemicals in the body lose an electron, they become positively charged (free radicals or oxidants). These chemicals roam freely throughout the body stealing electrons from other cells. Free radicals damage cellular DNA. Aging is the damage to millions of the body's cells through oxidation. This oxidation is due to the lack of available hydrogen anions to stop free radical damage.

Healing Power.

Molecular hydrogen is helpful for Rheumatoid arthritis (RA), a chronic inflammatory disease in which the progressive destruction of joints causes morbidity. It is also associated with an increased risk of atherosclerosis, resulting in cardiovascular disease and mortality. The therapeutic goal is to control systemic inflammation to reduce symptoms and improve one's general health status. Drinking water with a high concentration of molecular hydrogen significantly reduces oxidative stress in RA.[36]

Interestingly, water forms by combining oxygen (a powerful oxidizer) and hydrogen (a powerful reducer). It makes sense that molecular hydrogen has high-powered therapeutic potential, as does water itself. Hydrogen is a novel and innovative therapeutic tool. It is helpful like intravenous vitamin C therapy, except it is far less expensive and can be used around the clock. With an inhaler, one can pour hydrogen into the body directly through the lungs.

With every sip of hydrogen water, our bodies receive trillions of hydrogen molecules. H2 pairs up with toxic hydroxyl radicals to neutralize them. Hydrogen therapy is safe, and there is no upper limit of use.

Relative Sizes of Antioxidants

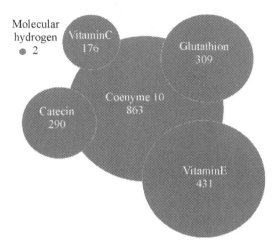

There is no toxicity to H2 because the by-product of the free-radical neutralizing reaction is water. Each molecule of H2 will neutralize two hydroxyl radicals into two molecules of H2O, hydrating your cells in the process. Hydrogen water at a concentration of 1.6 mg/L has more "antioxidant" molecules than 100 mg of vitamin C.

H2 is key to ATP production within the mitochondria. It is the mother of all other elements. Hydrogen, along with oxygen, has been intrinsically involved with the evolution of life. The extremes of oxygen and hydrogen provide a balance between oxidation and reduction, vital to life.

Hydrogen in the body is mainly bound to carbon, oxygen, and nitrogen. It is part of almost every molecule in your body: DNA, proteins, sugars, and fats. The hydrogen bond - which forms between atoms that "share" a hydrogen atom - is one of the most important interactions that make biological molecules behave as they do. Thus, hydrogen is essential in the regulation of physiology.

H2 is probably the only antioxidant molecule that can reach inside the mitochondria because of its small size. H2 directly protects mitochondria exposed to reactive oxygen.

The nasty hydroxyl radical has an unpaired electron, which turns it into an insatiable whirling dervish. Robert Slovak, a hydrogen water innovator, says, "It will steal an electron from DNA, cell walls, the mitochondria—and it will damage those when it does." The anti-oxidative stress effect of hydrogen happens by the direct elimination of hydroxyl radical and peroxynitrite. Subsequent studies indicate that hydrogen activates the Nrf2-Keap1 system.[37]

Hydrogen, Inflammation, and Pain.

Cells repeatedly exposed to inflammatory mediators will have genes that code for inflammatory response proteins switched on and thus be in a constant inflammation state. This leads to diseases and health conditions strongly linked to inflammation, including asthma and cardiovascular disease. Some of the molecules that can instigate a change in gene expression are Nf-kB, TNFa, and reactive nitrogen species such as nitric oxide and peroxynitrite. Hydrogen indirectly affects gene expression through its ability to modulate molecules that have a direct epigenetic effect. Molecular Hydrogen can:

- Impede release of NF-kB.
- Reduce TNFa.
- Reduce excess nitric oxide.
- Scavenge peroxynitrite.

Molecular hydrogen is an epigenetic modifying agent for genes that code for chronic pain and inflammation.

Therapeutic effects of molecular hydrogen for a wide range of diseases have been investigated since 2007. Most studies are from Japan, China, and the USA. About three-quarters of the articles show the effects in mice and rats with increasing clinical trials every year. One should note that almost all the initial research and publications came from Dr. Patrick Flanagan, developer of Mega hydrate. He was ahead of the pack in seeing how hydrogen can affect our health.

Hydrogen and Stress.

Emotions and stress matter very much in health and medicine. Few doctors are trained to help their patients with these issues. Hydrogen therapy combined with magnesium therapy helps reduce the price we pay on a cellular level for the stress we carry around. It is not just the stress from increasing background radiation, Wi-Fi and cell towers, pesticide residues in our food, and God knows what else that tears our cells down. It is also the emotional stress and feeling world of people that are all messed up.

All of this is more pronounced in cities where air pollution is damaging every organ. Air pollution damages virtually every cell in the human body. The research shows head-to-toe harm, from heart and lung disease to diabetes and dementia, and from liver problems and bladder cancer to brittle bones and damaged skin. Pollutants cause inflammation that then floods through the body, and nanoparticles being carried around by the bloodstream.

Usually, cells die through apoptosis. Necrosis is the death of body tissue. It occurs when too little blood flows to the tissue from injury, radiation, or chemicals. Pyroptosis is a form of programmed cell death that occurs during intracellular pathogen infection. Part of the body's antimicrobial response. The adaptive capacity of a cell ultimately determines its fate when it comes to stress. Strength, or what we call a cell's adaptive ability, is directly related to nutritional sufficiency and proper cellular respiration, removing toxins and wastes through the cell wall.

Cells can respond to stress in various ways, ranging from activating survival pathways to cell death initiation that eventually eliminates damaged cells. Whether cells mount a protective or destructive stress response depends on many factors, but the greatest has to do with nutritional status. A healthy cell naturally chooses life, but one that is already chronically stressed due to mineral and lipid deficiencies will have significantly less coping power, less resistance to stress. Other factors in calculating cellular resistance to stress are the nature and duration of the stress and the type of affected cells.

The cell's initial response to a stressful stimulus is to defend against, and recover from, the insult to it. So, it behooves us to maximize the cell's initial defensive response. Both hydrogen and magnesium will do this.

Magnesium and Cell Stress.

The involvement of free radicals in tissue injury, induced by magnesium deficiency, causes an accumulation of oxidative products in the heart, liver, kidney, skeletal muscle tissues, and red blood cells. Magnesium is a crucial factor in the natural self-cleansing and detoxification responses of the body. Magnesium protects cells from aluminum, mercury, lead, cadmium, beryllium, and nickel. This explains why re-mineralization is so essential for heavy metal detoxification and chelation. Magnesium protects the cell against oxy-radical damage and assists in the absorption and metabolism of B vitamins, vitamin C, and E - important in cell protection.

Glutathione, one of the essential enzymes in the body, requires magnesium for its synthesis. Magnesium deficiency causes glutathione loss. This loss is not affordable because glutathione helps defend the body against damage from cigarette smoking, exposure to radiation, cancer chemotherapy, alcohol, and just about every other kind of toxin. According to Dr. Russell Blaylock, low magnesium is associated with the increased free radical generation and glutathione depletion. This is vital since glutathione is one of the few antioxidant molecules known to neutralize mercury.

Magnesium is correctly called "The beautiful metal" by the ancient Chinese, for from a molecular biology point of view, the metal is priceless. Most of the hydrogen tablets that make hydrogen water use magnesium as their most active ingredient. In the best pills, one gets 80 milligrams, so if we take ten tabs a day, we get 800 mg, which is a healthy amount. This book on hydrogen and other medical gases does not pull into the magnesium port of wisdom overly. Still, it is evident: when employing medical gases, magnesium should always be included. (For further information about magnesium, please see my book *Transdermal Magnesium Therapy*.)

Oncologists need to know that magnesium also has a hand in protecting DNA, and it is a crucial ion in cell division. Likewise, pH control is key to cellular survival and determines much of a cell's adaptive capacity relative to stress. Iodine, sulfur, selenium, and even zinc are not far behind magnesium due to increasing cell strength. Bottom line: Hydrogen and magnesium are like Batman and Robin, inseparable medical superheroes for cell stress and just about everything else.

Hydrogen is a Medicine.

From the Department of Emergency and Critical Care Medicine at Keio University Hospital, Prof. Masaru Suzuki said, "H2 will become a medicinal product with a big impact in the field of cardiopulmonary resuscitation. Research is moving fast across the world, and it will eventually become a combined international investigation. If this H2 inhalation therapy works with cardiopulmonary arrest, it would mean that this treatment is effective even under the most severe conditions. I think the potential for H2 in medical uses will spread endlessly on the back of this medical research." [38]

In a study published in the Journal of Stroke and Cerebrovascular Diseases, Hydrogen Gas Inhalation Treatment in Acute Cerebral Infarction was safe and effective. These results suggested a potential for widespread and general application of H2 gas.[39] The medicinal value of hydrogen was ignored before research illustrating that inhalation of 2% H2 can significantly decrease the damage of cerebral ischemia/ reperfusion caused by oxidative stress via selective elimination of hydroxyl freebase (OH) and peroxynitrite anions (ONOO-). Hydrogen promotes the survival of retinal cells. One study demonstrated post-conditioning with inhaled high-dose H2 to confer neuroprotection against retinal I/R injury via anti-oxidative, anti-inflammatory, and anti-apoptosis pathways.[40] Hydrogen is a miracle when the only thing your eye doctor can suggest is antioxidants, and you are going blind.

> "Yesterday I went to my eye doctor. I told her about molecular hydrogen a few months ago and the relationship between oxidative stresses, the hydroxyl radical, age-related macular degeneration and glaucoma, and cataracts. Yesterday she compared pictures of my retina and was amazed. Not being knowledgeable about the eye, I did not understand the terms she used, but I could see that the retina surface was smooth now without an uneven surface. One obvious bump had disappeared totally. She wants me to come back in another month as part of a study she wants to write about," writes Bill.

Ed Wonder, one of the manufactures of hydrogen equipment, writes, "As you age, your ability to see up-close diminishes because of lens hardening. However, I was at the point where I couldn't focus because it was too close - and then when I moved it outward to get it into a focus-able range, it was too small—catch-22 on the vision. Also, reading around a bottle was difficult because of the different planes of the distance involved. That is gone, and I can read around bottles, at a comfortable distance, without the Catch-22 mentioned earlier. After four weeks of inhalation, my wife was able to get rid of her reading glasses completely, as well,"

Visual acuity improved after 3-4 weeks of daily inhalation. (Anonymous)

"I did some testing with my vision. I can read everything. I haven't seen this level since I was in my late 20's or early 30's. I am 53 right now. This is amazing.

"The ONLY thing that I changed in my life is the daily inhalation of the gas. The bubbled water did NOT do this, as I have been drinking that daily for six months - and only recently did things improve to this level. I have been doing 1/2 hour to 2+ hours daily.

We should only hope that hydrogen and oxygen work so well for us. We will see that hydrogen is effective in its power to bring people back from the edge of death like magnesium does when a person is experiencing cardiac arrest. What magnesium does for the heart, hydrogen can do, and together, of course, we have even more power over life and death.

Drs. Sun, Ohta, and Nakao, in their book *Hydrogen Molecular Biology and Medicine*, the first of its kind, write, "Inert gases are not useless or dispensable for human survival; instead, they are indispensable components of the gas medium to maintain life. To maintain the oxidation–phosphorylation energy metabolism, which is essential to life, the inhaled gas must contain a certain proportion of oxygen. But if the partial pressure of oxygen is too high or even pure oxygen, then it will cause damage to the body or even death since the presence of excess oxygen is toxic to the body."

"According to the research, hydrogen shows a protective effect in multiple diseases. For instance, malignant carcinoma, colitis, encephalopathy after carbon monoxide poisoning, cerebral ischemia, senile dementia, Parkinson's disease, depression, spinal injury, skin allergy, diabetes type 2, acute pancreatitis, organ transplantation, intestinal ischemia, systematic inflammation reaction, radioactive injury, retina injury, deafness, etc."

These doctors affirm that hydrogen gas's permeability is extreme because hydrogen gas's molecular weight is low. "It can penetrate rubber and latex tube at room temperature and can penetrate metal films such as palladium, nickel, and steel at higher temperatures." So, we can imagine hydrogen spreading throughout your body, getting to every cell, and especially to those areas that most need the healing power of hydrogen.

The more challenging the medical circumstance, the more helpful hydrogen seems to be. Hydrogen is perfectly suitable for intensive care wards (ICUs) and emergency rooms where medicines must act fast. According to the above authors, "The acoustic speed of hydrogen is fast. Under standard conditions, the acoustic speed of air is 331 m/s, helium's acoustic speed is 972 m/s, while the acoustic speed of hydrogen is 1286 m/s."

Molecular hydrogen contains two protons and two electrons, is neutral, so it does not take up as much space electromagnetically as negatively charged electrons. Meaning molecular hydrogen slips into cells more easily than the much smaller electrons. Cells resist any charge but not neutral hydrogen., So the tiny neutral molecules of hydrogen get into the cell and mitochondria with ease and speed, where they do a lot of good.

Hydrogen molecules quickly come to the rescue, riding on ultra-fast horses. Hydrogen is responsive; it is chemically active and energetic and will move like lightning in emergencies, especially when one combines inhaled hydrogen gas with hydrogenated water. The only thing more powerful than hydrogen water would be magnesium bicarbonate water.

Digging Deep into Pathology and Oxidative Stress.

Later in this book, we will discuss the nature of oxidative stress, why it is so damaging, and why hydrogen is the best solution to oxidative fire. Though I have not seen any studies directly on hydrogen and hypertension, the great silent killer, we read that hypertension is a multifactorial disorder that involves many mechanisms leading to risk factors for cardiovascular diseases.

Endothelial dysfunction is the imbalance between the production and bioavailability of endothelium-derived relaxing factors (EDRFs) and endothelium-derived contractile factors (EDCFs), associated with increased bioavailability of oxygen reactive species (ROS) and decreased antioxidant capacity characterized as oxidative stress. Hydrogen increases antioxidant capacity; thus, it should be seen as a basic solution to hypertension.

Hydrogen and Stroke.

Like natural gas, hydrogen can diffuse freely across biological membranes, acting in various functional capacities (Huang et al., 2010). Stroke is a devastating neurological disease. By inhalation, hydrogen gas can effectively pass the blood-brain barrier, leading to improved neurological deficits in various stroke models (Ohsawa et al., 2007; Chen et al., 2010; Lekic et al., 2011; Zhan et al., 2012).

Hydrogen Protects Against Pulmonary Hypertension.

Pulmonary hypertension (PH) is a condition caused by increased pressure in the pulmonary arteries. In advanced cases, its symptoms (shortness of breath, tiredness, chest pain) worsen and limit physical activity. Many factors account for the disease, and the most-used therapies rely on vasodilators of several kinds. However, traditional treatments have failed to block disease progress effectively. Thus, a search for new therapies is understandable.

Patients with pulmonary arterial hypertension (PAH) are at higher risk after all types of cardiac surgical procedures. These patients have higher incidences of hemodynamic instability, hypoxemia, and right ventricular dysfunction. Researchers have seen that H2 prevented the development of pulmonary hypertension and reversed right ventricular hypertrophy, a heart disorder characterized by thickening of the right ventricle walls.

Since oxidative stress and inflammation contribute to pulmonary hypertension's pathogenesis and development, we can understand why hydrogen would be helpful for this disorder.[41] One study, published in The Journal of Thoracic and Cardiovascular Surgery by Kishimoto et al., demonstrated that molecular hydrogen could decrease pulmonary arterial hypertension in a rat model by suppressing macrophage accumulation, reducing oxidative stress, and modulating the STAT3/NFAT axis.

In this study, pulmonary arterial hypertension was induced by injecting rats with monocrotaline. In the group treated with oral molecular hydrogen, H2 treatments decreased right ventricular systolic pressure, reduced the percentage of muscularized pulmonary arteries, and inhibited smooth muscle proliferation and vascular remodeling after 16 days.[42]

Importantly, medical scientists have seen that high doses of nonselective antioxidants do not increase the risk of morbidity and mortality. Therefore, selective antioxidants, like molecular hydrogen, are safer and more efficient therapy for PH patients. Molecular hydrogen (H2) is a particular antioxidant because it selectively reduces two specific ROS (hydroxyl radicals and peroxynitrite) without affecting beneficial and necessary reactive species necessary for cells' survival.

Hypertension.

Hypertension and diabetes are the principal cause of chronic kidney disease (CKD). Diabetes and hypertension affect up to 30% of the US population, so it is crucial to consider hydrogen to avoid these diseases. H2 gas has significant effects on the reduction of blood pressure without side effects. One study[43] aimed to investigate hydrogen (H2) gas on hypertension induced by intermittent hypoxia in rats. They exposed adult rats to chronic intermittent hypoxia (CIH) 8 hours/day for five weeks and H2 gas 2 hours/day. H2 gas significantly improved vascular relaxation.

Cannabinoids Reduce High Blood Pressure.

The body's endocannabinoid system plays a vital role in regulating many of its key physiological functions, including cardiovascular function. Anandamide – the body's naturally occurring THC version – relaxes blood vessels, allowing blood to flow more freely and lowering blood pressure.

Thus cannabis has a vasodilatory effect, which means that veins expand when you take marijuana. A clear sign is when you look into someone's eyes after they have consumed cannabis. The veins expand, allowing them to carry more blood at the same time. This is what causes the classic, red "stoner eyes."

After injecting rat subjects with THC, researchers found that blood pressure decreased significantly straight after. The same results were achieved in a second study. It is easy enough to find out if it will work for you. Take a few puffs. Measure your pressure before and after, and you will have your answer.

What's New in Hydrogen Medicine.

New hydrogen research is coming out all the time. One recent study gives weight to why I am almost always recommending hydrogen inhalation machines that add oxygen to the mix. When Hydrogen and Oxygen are mixed, improvements are seen in Cardiac Dysfunction and Myocardial Pathological.

During low oxygen/reoxygenation induced by CIH, a large number of reactive oxygen species (ROS) are generated and trigger oxidative stress damage. H2 quickly penetrates cell membranes and barriers without affecting basic metabolism in cells.

Results showed that the H2-O2 mixture remarkably improved cardiac dysfunction and myocardial fibrosis. We found that H2-O2 mixture inhalation declined ER stress-induced apoptosis via three major response pathways. The H2-O2 combination considerably decreased ROS levels via upregulating superoxide dismutase (SOD) and glutathione (GSH) and downregulates NADPH oxidase (NOX 2) expression in the hearts of CIH rats.

Treatment of Sleeping Disorders.

Obstructive sleep apnea (OSA) is a common breathing disorder characterized by recurrent upper airway obstruction episodes during sleep. The incidence of OSA is approximately 15-24% in adults. OSA is often accompanied by multiple cardiovascular disorders, such as hypertension, heart failure, and atherosclerosis. OSA patients showed long-term arterial oxygen saturation fluctuations and frequent sleep apnea, exposing them to a specific internal environment with chronic intermittent hypoxia (CIH) and recurrent hypoxia. Hydrogen and oxygen inhalation administered all night is a powerful intervention for treating intermittent hypoxia. Such treatment is what one would expect in ICU. One can apply such innovative Intensive Care at home.

Treatment of Hearing Loss from Chemotherapy.

Permanent hearing loss and tinnitus as side-effects from treatment with the anticancer drug cisplatin is a clinical problem. A preclinical in vivo study explored the protective efficacy of hydrogen (H2) inhalation on ototoxicity induced by intravenous cisplatin. Ototoxicity is being toxic to the ear (oto-), specifically the cochlea or auditory nerve, and sometimes the vestibular system, for example, as a side effect of a drug. The results of ototoxicity can be reversible and temporary or irreversible and permanent. The study concluded that H2 inhalation could reduce cisplatin-induced ototoxicity on functional, cellular, and subcellular levels. H2 inhalation mitigates the cisplatin-induced electrophysiological threshold shifts, hair cell loss, and reduced synaptophysin immunoreactivity in the synapse area around the IHCs and OHCs.

Treatment for Influenza.

Recent studies revealed that intraperitoneal injection of hydrogen-rich saline has surprising anti-inflammation, anti-oxidant, anti-apoptosis effects, and protected organisms against polymicrobial sepsis injury, acute peritonitis injury both by reducing oxidative stress and via decreasing mass pro-inflammatory responses. It is also well known that most viral-induced tissue damage and discomfort are mainly caused by an inflammatory cytokine storm and oxidative stress rather than by the virus itself.

Studies have shown that suppressing the cytokine storm and reducing oxidative stress can significantly alleviate the symptoms of influenza and other severe viral infectious diseases. Thus, medical scientists hypothesize that hydrogen-rich solution therapy is a safe, reliable, and effective treatment for Multiple Organ Dysfunction Syndrome (MODS) induced by influenza and other viral infectious diseases.

Conclusion.

Hydrogen gas inhalation is a powerful yet gentle therapeutic process. Conditions improve faster if hydrogen and oxygen are fed continuously. Hydrogen is the ultimate warrior against death and disease and is a perfect partner with oxygen, an ideal partner with CO2.

The Medical Context
of this Book

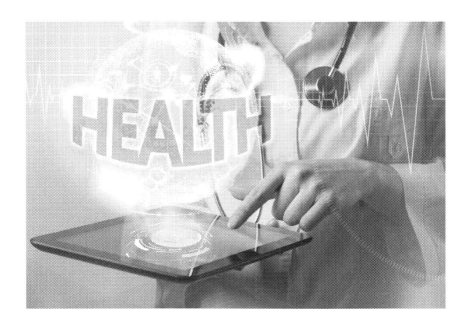

What is being offered is a precise form of medicine targeting the foundations of human life. This book is appropriate for physicians, nurses, alternative practitioners, and patients who will learn enough to stand on their own two feet to administer their treatments at home.

To give meaning to the complete system of medicine you will find in this book, I call my work Natural Allopathic Medicine, an approach that focuses on identifying and addressing the root causes of disease that are universal to almost everyone. Healing and medical science have become so confusing that practical measures can escape even the most educated physician. For many years I called my approach Natural Allopathic Medicine because it is based on western medical science but uses only concentrated natural medicines, many of which are used in ICU and emergency departments.

Natural Allopathic Medicine addresses the basic causes common to most patients, like low oxygen and carbon dioxide levels, acid pH conditions, mineral deficiencies, cold body temperature, breathing rate, and heart rate variability abnormalities.

A disease is usually the result of more than one cause, with the principal disturbances being in the most prime elements of life. To ignore the fundamentals is to ignore life and the most basic causes of disease. When an element fundamental to life goes lacking (becomes deficient), only supplementation and treatment with that essential element will suffice. Pharmaceuticals usually have the effect of further depressing essential elements making the patient worse instead of better.

Our approach gives patients and healthcare practitioners alike the tools to understand what is going on and what to do about it. Treatment protocols do not have to be developed based on repeated laboratory and clinical assessments, which all too often do not result in effective patient outcomes. Just ask anyone who has had to go from one doctor to another and get test after test without finding conclusive diagnosis and treatments that eliminate a person's presenting condition.

Instead of requiring a detailed understanding of each patient's genetic, biochemical, and lifestyle factors, we cover the fundamentals. We see what progress can be made without resorting to pharmaceutical medicines, surgery, chemotherapy, radiation treatments, and diagnostic tests that use cancer-causing radiation.

Of course, professional help can take us deeper into understanding ourselves if it is the right kind, but professional service can sometimes kill us when it's of the wrong type.

Though precise manifestation of causes in each patient depends on the individual's genes, environment, and lifestyle, only treatments that address causes will have lasting benefits beyond symptom suppression. Today's most advanced medicine forms are based on evolving research in nutritional science, genomics, and epigenetics. Still, we need to combine psychology and emotional medicine to round out our views. We humans are more than just our bodies.

By addressing the root causes, patients can act instead of getting swallowed up by Western diagnosis's seeming complexity by addressing the root causes rather than symptoms. For instance, after applying a complete protocol to address overall physiology, breast cancer patients are taught to apply topical treatments directly to the breasts using magnesium chloride, iodine, medical marijuana salve, and even clay to absorb toxins from the local tissues. Lung patients will be doing much the same using nebulization.

What is Innovative Intensive Care?

Every doctor who works with critical care knows magnesium chloride, sodium bicarbonate, iodine, potassium, and even injectable selenium. These primary minerals can be used to save lives in hospital settings and safely at home. Sodium bicarbonate, potassium chloride, and calcium chloride maintain pH and electrolytes within typical values in intensive care units. Magnesium is an essential mineral medicine for heart conditions and stroke. When it comes to sepsis, intravenous vitamin C and hydrogen gas inhalation therapy can save the day.

Vitamin C is an important independent antioxidant with a critical role in protecting cells from an oxidative challenge and cell death from oxidative stress. When used in combination with other antioxidants like hydrogen, glutathione, magnesium, bicarbonate, and iodine, one can feel confident that such a protocol will address sepsis's most serious threats as other life-threatening diseases.

These vital natural medicines can be used at home many times a day at high doses to effect significant changes in a person's medical situation safely, even if they are at or near death's door. One can apply lifesaving medicinal around the clock at home orally, transdermally (topically), and via nebulization, enemas, feeding tubes, baths, and even intravenous methods if a nurse is available. These substances all heal through the fulfillment of nutritional law.

These are the kinds of medicines that address the root causes of disease, those causes that are fundamentally the same for everyone, no matter what the symptoms or how each person's disease manifests.

Clear Medical Thinking.

Innovative medicine restores the capacity to think in clear medical terms. For example, it should be evident that the only cure for magnesium deficiency diseases is magnesium. Magnesium deficiency underpins many diseases, yet doctors rarely prescribe magnesium as a medicine. Millions die from cardiac arrest that could be avoided if magnesium would have been prescribed as an FDA-approved pharmaceutical treatment. The only problem is that there is no money in it. The same is true for bicarbonate, so patients do not hear about either from their doctors.

This book introduces a form of medicine grounded in essential physiology. In common sense, one does not need a Ph.D. to understand. It is a book of clear medical thinking, which has been lost through the past century as medical belief systems motivated by profit took root.

This book is an excellent companion to Conquering Cancer-a cause in Naturolpathic Oncology, which goes much deeper into oxygen and carbon dioxide medicine. It is essential to know from the beginning that oxygen is dangerous without CO2, and we do have CO2 deficiencies occurring in the blood and cells all the time. Plants thrive on higher CO2 levels, and so do we.

The main point of this book is to promote hydrogen therapy. Anyone with a life-threatening disease needs hydrogen inhalation therapy. However, anyone who needs hydrogen therapy also needs oxygen, and that is the primary reason I recommend hydrogen inhalation devices that output both gases. This book is about leaping tall medical buildings in a single bound with the three primary medical gases.

The basement of Life.

Inflammation is inseparable from lower pH, oxygen, and CO2 deficiencies, tracking body temperature, respiration, and elimination. A unification theory of medicine would describe this area of physiology where things are happening simultaneously.

There is a point where one cannot separate oxygen from CO2 levels because they are locked into a tight mathematical relationship. The same about pH and cell voltage. As CO2 levels go south, so do O2 levels; when pH dives, so does cell voltage.

Cancer.

Given enough time, cancer will develop whenever there is a proliferation of damaged cells, and when oxygen levels fall, cells are damaged, cell wall permeability changes. When toxins and free radicals build up, and the mitochondria lose functionality in terms of ATP production, and when pH shifts to acidic in cells, when essential gasses like carbon dioxide are not present in sufficient concentrations, and when essential nutrients are absent, if stress levels are high enough, cells will eventually decline into a cancerous condition.

What is the Truth about Medicine?

Mainstream medicine is not kind to natural substances, giving them bad press even though they save lives in dire medical situations. Even Google is getting into the act. Google states that "pages that directly contradict well established scientific or medical consensus for queries seeking scientific or medical information" are suppressed on its search engine, "unless the query indicates the user is seeking an alternative viewpoint."

"Facebook has deleted dozens of pages dedicated to fringe or holistic medicine in an apparent crackdown on pseudoscience. The Global Freedom Movement, an alternative media site, reported that the social platform purged over 80 accounts and that "no reason was provided. No responses to inquiries have been forthcoming."

I wonder what people think about antibiotics today. Soon, if not already, it will become too dangerous to step foot in a hospital because antibiotic-resistant infections will run rampant, leaving doctors impotent to treat the most common pathogens.

Violent Medicine.

The greatest proponent of vaccines, Dr. Paul Ofitt, tells us that administering vaccines is a violent act.

> "Vaccinations are not easy. We ask a lot of our citizens to get as many as 26 inoculations in the first few years of life, five shots at one time. It's hard to do that, especially given that vaccination is a violent act; you pin the child down, you give them this biological agent against their will."

The medical-industrial complex will poison more people without open debate without hearing contrasting views, which is in love with the 'dose makes the poison' paradigm. Hydrogen is not a poison. Neither is oxygen when balanced with CO2 and enough hydrogen to keep oxidation running cool (without excess oxidative stress).

I am proud to redefine medicine into something vastly safer and more effective. There is no point in being nice about what is going on in the world of medicine, held captive by the pseudoscience of pharmaceutical research. It is too much for doctors to understand that pharmaceutical medications' toxic side effects are their main effects. No matter how low the dose—a poison is still a poison.

Hydrogen for Surgery & Intensive Care.

Medical gas is critical to the function of hospitals and many other healthcare facilities. Medical gas systems in hospitals are, in a word, lifesaving. Piped in medical gases, oxygen, nitrous oxide, nitrogen, carbon dioxide to hospital areas such as patient rooms, recovery areas, operating rooms, and ICU departments is critical to patients' survival. Now hydrogen needs to be added to the list.

Hospitals must invest in hydrogen because it is a perfect and safe substance for oxidative stress. Evidence of massive oxidative stress is well established in adult critical illnesses, characterized by tissue ischemia-reperfusion injury and an intense systemic inflammatory response such as sepsis and acute respiratory distress syndrome. Oxidative stress exacerbates organ injury and the overall clinical outcome.[44] Oxygen-derived free radicals play an essential role in the development of disease in critically ill patients.

"Critically ill patients suffer from oxidative stress caused by reactive oxygen species (ROS) and reactive nitrogen species (RNS). Although ROS/RNS are constantly produced under normal circumstances, critical illness drastically increases their production. These patients have reduced plasma and intracellular levels of antioxidants and free-electron scavengers or cofactors. They also have decreased activity of the enzymatic system involved in ROS detoxification. The pro-oxidant-antioxidant balance is of functional relevance during critical illness because it is involved in the pathogenesis of multiple organ failure."[45] Hydrogen is the gas that directly and immediately addresses critical conditions resulting from massive oxidative stress.

Various studies suggested the therapeutic effects of hydrogen gas concerning multiple aspects of emergency and critical care medicine, including acute myocardial infarction, cardiopulmonary arrest syndrome, contrast-induced acute kidney injury, and hemorrhagic shock.

There are emergency room / ICU / Operation room hydrogen machines with a Gas Production Rate of Hydrogen /Oxygen: 3000ml~6000ml/ min. Hospitals and the FDA officials might not know it yet, but China already produces hydrogen machines perfected to the ICU and operating rooms' climate. Hospitals can afford the 30,000 dollars estimated price and absolute control over the combined gases. Less costly units will flood the body equally well with hydrogen in your home.

Hydrogen works fast because of its small size and neutral footprint. Inhalation of 1–4% hydrogen gas alleviated tissue damage and reduces infarct size. The blood and tissue levels of hydrogen reach saturation within 20 minutes after commencing hydrogen gas's inhalation. The gaseous hydrogen level in the blood reaches 16 μmol/L after inhalation of 2% hydrogen gas. Arterial oxygen saturation is not affected because gaseous hydrogen does not bind with hemoglobin, and the blood pressure and pulse rate are also unaffected under steady-state conditions. The blood level of hydrogen gas declines rapidly after discontinuation of inhalation because the lungs excrete it."[46]

Once the blood is saturated, it is only a matter of keeping hydrogen levels available for circulation. Hydrogen inhalers using 99 percent hydrogen will saturate the body even quicker.

Hydrogen Science.

In a rat model of resuscitated cardiac arrest, the survival rate at 72 h after the return of spontaneous circulation (ROSC) was only 30% in the control group. In contrast, it increased to 70% in the hyperthermia group and the hydrogen gas group and was even higher at 80% in the combined group.[47] Combining hydrogen with infrared warming therapies optimizes therapeutic outcomes.

Stroke is a devastating neurological disease, and hydrogen has shown promise for these patients as well. By inhalation, hydrogen gas can effectively pass the blood-brain barrier, leading to improved neurological conditions in various stroke models (Ohsawa et al., 2007; Chen et al., 2010; Lekic et al., 2011; Zhan et al., 2012).

"Hydrogen gas inhalation confers resistance to hemodynamic instability caused by massive bleeding. Breathing hydrogen gas when homeostasis is disturbed, it works on complex networks and restores homeostasis. Endogenous physiologically active gases, such as nitric oxide and carbon monoxide, bind to heme, but hydrogen gas does not."[48]

Safer Surgery with Magnesium.

Hydrogen works well with magnesium, and so does sodium bicarbonate. In surgery, emergency room medicine, and intensive care, magnesium is present. Complications such as arrhythmias, kidney failure, stroke, and infections may occur after major surgery. Everyone scheduled for surgery needs to increase their stores of magnesium. In the before and postoperative phases, magnesium can help alleviate pain, decrease blood pressure, alleviate certain heart arrhythmias, prevent blood clotting, relieve depression so common after bypass surgery, and improve energy and cognitive abilities.

When magnesium levels are corrected before, during, and after surgery, medical complications are reduced to the point where it becomes merely imprudent to perform surgery without it. Dr. Minato at the Department of Thoracic and Cardiovascular Surgery in Japan strongly recommends the correction of hypomagnesemia during and after off-pump coronary artery bypass grafting (OPCAB) prevention of perioperative coronary artery spasm. His team has said that they will no longer perform this surgery without its use.[49]

Conclusion.

Hydrogen is another substance that will facilitate positive outcomes in ICU, surgery departments, ambulances, and emergency centers. In the future, it will be a grave error, even malpractice, not to give hydrogen or magnesium in these critical care departments. It should not be a great leap for the medical system to provide oxygen in all instances to giving hydrogen and oxygen and magnesium at the same time.

The high-end hydrogen inhaler featured above would give doctors perfect control over both gases. However, for the patient, hydrogen inhalation needs to be used at home daily to receive the optimum benefit. Personally, except for surgery, I see no reason for such expensive equipment. Other inhalers more economically put out enough healing gas to level the playing field between hydrogen inhalers.

Saving a Life with Hydrogen and Oxygen.

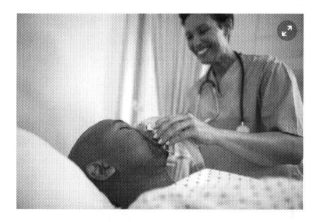

After receiving an email that someone was in danger of death, a friend offered a hydrogen-oxygen (HydrOxy) gas generator. This man spent 9+ hours inhaling HydrOxy gas through a nasal cannula –and the most remarkable changes transpired during that time.

Upon arrival, he did appear to be close to death. He was ashen, largely unresponsive (in and out of consciousness), mouth-breathing, had started to have difficulty swallowing (a common occurrence with dying patients), and was not eating anything. His wife thought he wouldn't live another 24 hours. He was hooked up to the HydrOxy with the nasal cannula right away, although his mouth-breathing gave pause that he might not be getting the full benefit. He was encouraged to breathe in through his nose and out through his mouth as much as possible. It was not long before his breathing began to deepen. Gradually but perceptibly, the color started to return to his cheeks and forehead. His wife watched him very closely, although her emotional involvement may have kept her from observing the subtle changes.

After he needed to use the toilet, he propelled himself on a rolling chair to the bathroom, and she noticed the obvious and undeniable turnaround. She gave the technician who brought the hydrogen inhaler a big hug saying, "You saved his life!" He came out of the bathroom, wanting to go to bed, and proceeded to head up the stairs by himself. Knowing his weakened condition, she was concerned he might fall.

Once she got him into bed, she connected him to the hydrogen gas again. As he continued breathing with the nasal cannula, his improvement was that she felt comfortable asking the housekeeper to keep an eye on him while she went to get some supplies. It was apparent that he was not only doing well, but he had left the bed to use the bathroom again, and after returning, he had the presence of mind to re-install the nasal cannula himself.

Upon returning to the house, his skin color and tone – face, chest, and body – was markedly better, his extremities were warm, and he was pretty lucid, even kissed his wife. He could drink 8 oz. of soup broth, the first nourishment he had taken for a while. By now, his wife wanted to know if he could leave the inhaler on all night. There was no reason why not if his wife would be sleeping by his side.

Nothing in the medical literature advises against breathing this gas for prolonged periods. In medical studies conducted by doctors in Asia, Molecular Hydrogen seems to be one of those rare applications that fit the expression "more is better!" A wonderful thing happened that day. And thanks to a relatively unknown (in the United States) technology, this man was no longer at death's door — he was thriving.

Regrettably, this man passed away but lived another six months after receiving help from the HydrOxy, another name for Brown's Gas.

Saving Sepsis Patients' Lives.

Are doctors ready to save millions of people who die from sepsis? According to the National Institutes of Health, we now have a game-changer for a condition that occurs in more than 1.5 million people a year in America alone, with a 28-50% fatality rate. Sepsis death rates are even higher in third-world countries.

With sepsis, it's a matter of death for more than 500 thousand Americans a year. Hydrogen can save most of these patients, together with oxygen, Vitamin C, D, and CO2, administered with hydrocortisone and thiamine. Hydrogen would push survival rates higher with injectable selenium, IVs of sodium bicarbonate, and magnesium chloride.

Selenium is an excellent anti-oxidant anti-inflammatory, protecting against reperfusion injury, myocardial infarction, ischemic stroke, and vascular surgery alleviated with selenium injections, as would cytokine storms resulting from out-of-control infections.

> Science Daily reports, "With infectious diseases, it is often not the pathogen itself, but rather an excessive inflammatory immune response (sepsis) that contributes to the patient's death, for instance, because of organ damage. In intensive care units, sepsis is the second most common cause of death worldwide. In patients with a

severely compromised immune system, especially, life-threatening candida, fungal infections represent a high risk of sepsis."

Sepsis, a systematic inflammatory response to infection, is among the most severe diseases in an ICU. Even with comprehensive therapy, sepsis is still associated with high morbidity and mortality until Dr. Paul Marik came along.

Dr. Marik made headlines across the globe with an innovative sepsis treatment. However, he must "lie low" about the controversial treatment. The response by patients' physicians has been about half and half; some willing to try, and others were saying it's complete and utter nonsense." He is administering a common-sense basic medicine approach that puts out cysteine storms with IV infusions of vitamin C, hydrocortisone, and thiamine. [50]

His hospital experimented with his protocol and saw a stunning mortality drop. However, doctors are still skeptical because his study involved only a small number of subjects and the survey than an earlier control group. However, medical institutions are taking this seriously enough to launch more investigations.

Published in "Chest," an American College of Chest Physicians medical journal, a study of 47 patients with sepsis treated in Norfolk General's ICU in 2016, only four died, but of the conditions that led to sepsis in the first place. The previous year, 19 of the hospital's 47 septic patients died. Dr. Marik has treated 700 patients with the protocol, and while some have died, it has usually been because of the underlying disease, such as cancer, that led them to a septic state.

Dr. Marik did not use hydrogen in his therapy. But because molecular hydrogen is an antioxidant, more primordial than vitamin C, this chapter proposes to increase the effectiveness of Marik's novel treatment by adding hydrogen to the mix.

Vitamin C, Cancer and Oxidative Stress

A research team at the University of California, Berkeley, found that vitamin C significantly reduced oxidative stress levels associated with various chronic diseases for people exposed to environmental tobacco smoke. Vitamin C also lessens imidacloprid-induced oxidative damage by decreasing LPO (lipid peroxidation levels) and altering the liver's antioxidant defense systems.

Vitamin C is an essential anti-oxidant protecting cells from an oxidative challenge. High levels of hydrogen enhance ascorbic acid's anti-oxidizing activity threefold crucial in avoiding the pro-oxidant risk of a high dose of ascorbic acid.

Vitamin C is not without its limitations. "Destructive ROS like hydroxyl radicals are strong oxidants that cause tissue damage, whereas beneficial species like superoxide and hydrogen peroxide enhance endogenous anti-oxidant mechanisms through signal transduction pathways. A potent anti-oxidant, such as vitamin C, indiscriminately eliminates both destructive and beneficial ROS, thus failing to suppress the progression of conditions related to oxidative stress. Hydrogen gas is a weak reducing agent, and its oxidation-reduction reaction only occurs with a strong oxidant that causes tissue damage."[51]

Molecular Hydrogen

Molecular hydrogen will extend our stores of Vitamin C. Molecular hydrogen uplifts other antioxidant activity because an adequate supply of vitamin C enables the regeneration of vitamin E and other anti-oxidants in the body. IV vitamin C is a fair treatment for people who are close to death's door. It has the power to bring people back from the brink, and so does hydrogen.

Sepsis, a multiple organ dysfunction syndrome, is the leading cause of death in critically ill patients. Hydrogen gas inhalation significantly improved the survival rate and organ damage of septic mice in laboratory tests.

Biopsies proved hydrogen's protective effect on sepsis. Molecular hydrogen therapy can significantly reduce the release of inflammatory and oxidative stress injuries, reducing the damage of various organ functions common with sepsis.[52]

Hydrogen Coronavirus Therapy.

Dr. Cameron Kyle-Sidell, MD, is a board-certified emergency medicine physician in Brooklyn, New York. He is affiliated with Maimonides Medical Center and is crying out that something is not right in ICU departments across America regarding how they are treating coronavirus patients.

Dr. Isaac Solaimanzadeh, a practitioner of Internal Medicine at the Interfaith Medical Center in Brooklyn, supports what Dr. Kyle-Sidell says in a video about coronavirus being something more like high altitude pulmonary edema than a viral driven pneumonia. He stated:

> "While anti-viral approaches and vaccines are being considered, immediate countermeasures are unavailable. COVID-19 exhibits decreased arterial oxygen partial pressure to fractional inspired oxygen with attendant hypoxia and tachypnea. There also appears to be a tendency for low carbon dioxide levels in both as well."

Carbon dioxide is an essential medical gas. Medical gases trigger naturally occurring physiological responses, enhancing the human body's preventive and self-healing capabilities. Medical gases include carbon dioxide, oxygen, nitrogen, nitrogen oxide, and most recently, hydrogen can be employed at home or in ICU to treat a coronavirus infection's worst symptoms.

Hydrogen to the Rescue if You Cannot Breathe

Ford and GE Healthcare's goal was to produce 50K ventilators within 100 days. Everyone was scrambling for ventilators. But what they should have been scrambling was to get hydrogen inhalation devices. Too many people on ventilators died, yet people can quickly start breathing with ease again with hydrogen gas.

Coronavirus medical thinking is evolving quickly. Approaches can be combined into a complete protocol that will keep 99 percent alive. Since the virus that roared hit the streets of China, medical officials have not budged an inch. The official story remains the same. No treatment, no vaccine. Now we have experimental non-FDA approved vaccines, and as of the date of this publication, over 15 thousand deaths have been reported into the European and American vaccine reporting systems.

With hydrogen inhalation machines instead of ventilators, we would be witnessing an entirely different story. Lower death rates mean we would not have had to destroy the world's economy. We have every reason to start a Manhattan type of project and build thousands of hydrogen inhalation machines and not have to worry about coronaviruses killing us.

Prof. Zhong Nanshan, a Chinese epidemiologist and pulmonologist (discovered the SARS coronavirus in 2003), recommends the hydrogen-oxygen gas mix inhalation based on Chinese patients' data from Wuhan. National Health Commission of the People's Republic of China says, "Inhaling mixed 66.6% hydrogen and 33.3% oxygen to treat the covid-19 virus pneumonia." It is on the top of their treatment list for seriously ill patients.

Transcript:
Doctor: Tell me how you feel after inhaling hydrogen and oxygen gas?

Patient: After inhaling, I had been here for so many days; my chest pain had not been alleviated. Indeed, after I used Hydrogen-Oxygen Atomizer, my chest pain was relieved. No chest pain anymore, no pain at all.

Doctor: So great!

Patient: Thank you, Prof. Wei Chunhua. Today, here for the first time, I feel free now.

Transcript:

Doctor: How are you feeling?

Patient: With this Hydrogen Oxygen Atomizer, I feel fine.

Doctor: Do you feel that with this device, your breath is more comfortable?

Patient: Yes, it is.

Doctor: Your breath is smoother, isn't it?

Patient: Yes, it is. When I used it for the first time.

Doctor: How did you feel?

Patient: I felt as if the gas mask was removed.

Doctor: This feeling.

Patient: Yes, that. I dared not take a deep breath in the past, but I can take deep breaths with the Hydrogen – Oxygen.

Doctor: Good! Great!

Transcript:

Doctor: How are you feeling?

Patient: I'm feeling comfortable.

Doctor: Comfortable, yes?

Patient: Yes, nothing is wrong.

Doctor: Do you feel your chest is very relaxed after inhaling? Or do you feel your inhalation is a little smooth?

Patient: yes, I do. In the past, when I took a deep breath, I would cough. But with the Hydrogen Oxygen Atomizer, the cough almost disappears. And when I take a deep breath, I do not cough.

Doctor: Good, very good!

Doctor: What is your feeling now?

Patient: After breathing with it a half a day, I feel my chest does not pain anymore. Not pain so much, and now, I feel no chest pain at all. No pain now. Before this, my chest did pain. It was a chest tightness. Front, this place, and back ached a little. After I used this to breathe in hydrogen for half a day, I felt much better, with no pain. Good! Thank you.

In light of the significantly decreased airway resistance and safety, inhalation of hydrogen and oxygen mixed gas generated through water electrolysis was applied in ICU medicine. Hydrogen/Oxygen mixed gas inhalation significantly ameliorated dyspnea in most patients with Covid-19 in a pilot investigation. Therefore, it has been endorsed by the latest Recommendation for the Diagnosis and Management of Covid-19 document.

Brown's Gas is a Mixture of Hydrogen and Oxygen.

Brown's Gas has already been endorsed for COVID-19 therapy by the Chinese Government, EU organizations, and UK Doctors. Brown's Gas (aka HydrOxy or HHO) inhalation has helped treat pneumonia caused by a coronavirus. Hydrogen molecules do not directly target the new coronavirus but can eliminate the virus's inflammation to play an auxiliary therapeutic role. The most superior feature is anti-inflammatory without side effects, reported by the national health commission of China.

Hydrogen can both down-regulate expression of oxidative-related genes and pro-inflammatory cytokine genes, directly and indirectly. Oxidative stress and systemic inflammatory response syndrome play a critical role in tissue and organ damage after polymicrobial sepsis injury, acute peritonitis injury, and peritonitis, which can develop into lethal sepsis inappropriate treatment.

Until now, studies have indicated cytokine storm and oxidative stress are highly associated with a pathological process when getting infected with viruses. Although cytokine storms and oxidative stress try to eliminate these pathogens, they seem to generate multi-organ damage, resulting in lethal clinical symptoms such as extensive pulmonary edema, alveolar and other tissue hemorrhages, and acute respiratory distress syndrome, etc.

Nitric Oxide.

The same is being said about nitric oxide. In 1992, the journal Science named nitric oxide "molecule of the year." And in 1998, UCLA pharmacologist Louis J. Ignarro shared a Nobel Prize in medicine for uncovering nitric oxide's role as a "signaling molecule in the cardiovascular system." Nitric oxide is a gas; once inhaled, it works by relaxing smooth muscles to widen (dilate) blood vessels, especially in the lungs. Also, it is used together with a mechanical ventilator to treat respiratory failure in premature infants. Increased levels of CO2 also dilate blood vessels while positively affecting oxygen disassociation.

At hospitals in Boston, Alabama, Louisiana, Sweden, and Austria, researchers have launched a clinical trial to test inhaled nitric oxide in patients with mild to moderate cases of COVID-19. The trial will evaluate whether the gas can drive down the number of patients who end up needing breathing assistance from a mechanical ventilator. For about 30 minutes two or three times a day, study participants assigned to the trial's active arm will inhale a high dose of nitric oxide through a mask.

In Italy, where the gas was used under more haphazard conditions, the treatment appeared to boost oxygen levels in the blood of COVID-19 patients dramatically, said Dr. Lorenzo Berra, the critical-care specialist at Massachusetts General Hospital leading the new trial.

Flexibility of Hydrogen Administration.

Interestingly, H2 therapy can be administered through inhalation, oral intake of hydrogen-rich water, injection of hydrogen-rich saline, direct diffusion of hydrogen: bath, eye drops, and immersion, and increase hydrogen in the intestines. Although each delivery way has its characteristics and advantages, injection of hydrogen-rich saline allows enough hydrogen to have antioxidant, anti-inflammation, and anti-apoptosis effect in the shortest time.

Conclusion.

We need to listen to the best doctors who are on top of the situation, not the media and health officials making wildly different predictions.

It's *ridiculous* that the modern medical establishment won't talk about low-cost prescription drugs (like hydroxychloroquine) or simple nutritional supplements like zinc, magnesium, Vitamin C, D, and selenium that are widely available and have a long track record of safe, effective use. When concentrated, they all make excellent medicines in both emergency and ICU departments. They certainly will not talk about sodium bicarbonate even though we have both old and new indicators for its use.

As mentioned above, Dr. Isaac Solaimanzadeh tipped us off that COVID-19 tends to low carbon dioxide levels. How do we increase patients' carbon dioxide levels? Give bicarbonates in sufficient quantities orally or intravenously. Health officials are not doctors like those who are in the front lines treating desperate patients in ICU. In the context of this global pandemic, the active censorship of alternative treatments is a crime, a sadness, and a lost purpose to the fundamental mission of all practitioners of the healing arts.

Perhaps health officials will have to defend themselves to history for their courses of action that destroyed an entire civilization's way of life instead of embracing things that work.

Revolution in Intensive Care Medicine.

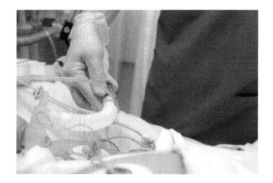

Inside emergency rooms and intensive care wards are common but extraordinarily safe and effective substances that save lives every day. Interesting that no one has thought to harness these super medical weapons against chronic disease or cancer.

In emergency rooms and intensive care wards, medicines must be safe while, at the same time, delivering an instant lifesaving burst of healing power. If they are safe and effective for emergencies, they will commonly face chronic and acute diseases because they address nutritional deficiencies.

What surprises most doctors and many patients is that the best healing agents are not pharmaceutical medicines but highly concentrated nutritional substances. These offer a power unequaled in the world of medicine. Medications that can save a life on demand instantly in an emergency, like cardiac arrest, are the same to battle diseases that have baffled the entire western Allopathic establishment for decades.

Because nutritional medicines are non-toxic, we can layer treatments and attack from many sides in a simultaneous assault on cancer or anything else threatening life or making us ill. Hydrogen, oxygen, sodium and potassium bicarbonate, magnesium, iodine, selenium, medical marijuana, oxygen, CO2, glutathione, Vitamins C and D, and sulfur are the principal natural agents used in innovative intensive care as well as in your own home.

Sodium bicarbonate, a standard emergency room medicine, acts as a powerful, natural, and safe anti-fungal agent. When combined with iodine, it covers the entire spectrum of microbial organisms. The efficacy of sodium bicarbonate against certain bacteria and fungi has been documented, but its role as a disinfectant against viruses is barely known. Sodium bicarbonate at concentrations of 5% and above is helpful with a 99.99% reduction in viral titers on food contact surfaces within a contact time of 1 min.[53]

Intensive Care for Stage Four and Five Cancer.

Hydrogen-driven protocols provide us with an entirely new concept in intensive cancer treatment. There are four stages of cancer, but when doctors give up on us and give us an imminent death sentence—we can consider this stage five. A stage five cancer demands emergency intensive care procedures to bring people back from death's door. Similar to stage four cancer, though there is a little more time.

The increasing prevalence of cancer patients leads to a growing number of cancer patients who will require intensive care treatment. Advances in critical care have led to increased survival of critically ill patients with cancer. Acute care is becoming an essential cornerstone of modern cancer care.

In intensive care wards, most medicines are given intravenously or through injection. Some people can do this at home as well. However, most of us can duplicate such administration's intensity through intensive medical baths, transdermal medicinal applications, nebulization, and oral intake by loading one's water with powerful medicine. We have home-use hydrogen and oxygen machines that give us concentrated power to heal the sick. The treatment has never been better.

One can learn to practice intensive care medicine in one's own home more safely and much less expensive than in a hospital.

Directions for use for emergency and intensive care.

I recommend connecting to a hydrogen inhaler for intensive care patients and leaving the gas flowing continuously until a noticeable change occurs. Expect such occurrences. Parallel to this is hydrogen water intake, even if given through a feeding tube.

Intense treatment should show measurable and noticeable results within eight hours to one to two days. If the hydrogen inhaler also mixes oxygen into the cannula, we must add magnesium, selenium, iodine, sulfur, and bicarbonate to the protocol for best results.

Special Note: Researchers report that breast cancer can "smolder" and return even 20 years later unless patients keep taking drugs with debilitating side effects to suppress it. It will not "smolder" if enough hydrogen is taken to make sure cellular "fires" do not lead to another cancer. Add to the list iodine for iodine deficiencies make the breasts exceptionally vulnerable to cancer.

Case Study – Salivary Cancer

This is about a man with salivary cancer, which had spread to include brain stem cancer, lung cancer with something not determined affecting the optic nerve. Everything crashed for him after the radiation therapy to the brain about 15 months ago, and he could not recover from that as we would expect. This past February, given three weeks to live diagnosis, the patient was very frail, bleeding under the skin like tissue paper. He could hardly walk (we would almost carry him), eyes closed and barely speaking, liver and kidneys shutting down, and unable to use his bowels on his own.

However, he was given hydrogen and oxygen gas therapy near the end and made considerable progress in February. His improvement was stunning. The therapists administering the hydrogen and oxygen gas therapy thought he would end up with a complete recovery. They watched the life return to him. A spark in his eye was the first thing they saw, which was on the first day. His skin thickened, and his bruising faded. His yellow look and smell returned to normal. His bowels started moving as they should. From there, his appetite returned, although putting on weight was still a problem as the radiation had broken someone who was a strong and fit man only 15 months before. His strength was better, and he could now walk to the toilet on his own without falling and cracking jokes and being cheeky with his wife. No wonder the therapists were optimistic.

All this reversed when the doctors changed his drug schedule. His death is now imminent. His wife did all she can, fighting for him for so long, but the system continued to force her hand every step of the way.

The people who reported this case fear she will deliberately follow him shortly after losing him. This one story of hydrogen and oxygen therapies, when combined, can do and how difficult things can be for terminal patients when there are still doctors involved.

Hydrogen Inhalation Devices.

Hydrogen inhalation devices are available in the United States and are already widely used in Japan and China. In the Hydrogen for Surgery & ICU chapter, we saw a high-end hydrogen oxygen inhaler for surgery and intensive care under development that will go for thirty thousand dollars when approved by the FDA. As you will see, you can get a good hydrogen inhalation machine for as little as 2,000 thousand dollars.

The average couple taking molecular hydrogen tablets consumes two bottles per month at an average cost of $119 for the highest parts per million hydrogen tablets on the market. When you own a hydrogen inhalation machine, the whole family can use it at a fraction of the cost vs. hydrogen tablets. For reference, just 40 minutes on a hydrogen inhalation machine is the equivalent of taking an entire bottle of Vital

Reaction Molecular Hydrogen Tablets! If you use your inhalation machine daily for 40 mins for a year, that is equal to over $21,700.00 worth of tablets!

This chapter is devoted to options and uses available for patients and clinics. There is a wide range of prices of Hydrogen Inhalers. The below units will deliver most, if not all, the safety and therapeutic benefits of a thirty-thousand-dollar machine but cost between two and seven and a half thousand dollars. My recommendation for most patients is to use a device that produces a mixture of hydrogen and oxygen called HydrOxy (aka Brown's Gas). HydrOxy is a mixture of 67 percent hydrogen and 33 percent oxygen. When water is turned into HydrOxy, it expands (from liquid to gas) 1866 times. One liter of water can make 1866 liters of HydrOxy. The unique properties of this gas and its therapeutic effects have been defined in advanced hydrogen research.

Most hydrogen inhalers will saturate the body quickly. Inhaled air containing hydrogen will reach a peak plasma level in about 20 min, and upon stopping of inhalation, the return to baseline takes about 30 min.

Inhalation is the most effective way to get hydrogen into the body. Inhaling air containing 2% hydrogen for 12 seconds will put as much hydrogen into the blood as drinking a liter of water infused with hydrogen at 1.6 ppm.

Note that all gas coming from a machine is not used because a breathing cycle is 1/3 inhalation, 1/3 exhalation, and 1/3 rest. So high output flow rate machines are not as efficient as companies make them out to be.

If used as directed and maintained at the recommended service schedule, the machines below are both expected to last well beyond ten years, even if used 24/7. All have power supplies compatible with 120 VAC and 240 VAC.

Below is a testimony I received recently from a remarkable woman who runs a healing retreat center in Texas and Costa Rica. She shared with me her clinical experience that suggests that hydrogen inhalation (hydrogen and oxygen mixture called Brown's Gas) is the most effective, fastest, and safest way to eliminate inflammation and that it is also the most adjuvant treatment for every condition.

Genita M. Mason, Medical Director at The Ozone Treatment Center in Texas, said," through these months of witnessing both myself and my patients' excellent results in all conditions treated, I can honestly say that adding brown gas hydrogen (66% hydrogen and 33% oxygen) using the Hydrogen Technologies inhalation machine to my already high impact biological medicine model has blown our already superior patient outcomes through the roof! It is also important to note that the 66H2/33O2 mixture is the best for patient recovery outcomes."

"My patients and I have experienced what I would call miraculous results if I did not understand the science behind what is happening "under the hood." Fatigue, brain fog, visual issues, depression, anxiety, low hormones, athletic endurance, neurodegenerative disease, cancer, cardiovascular disease, diabetes, biotoxin illness, all gut issues – essentially every condition we have treated after implementing hydrogen inhalation and water to the program has dramatically improved or completely cleared," wrote Mason.

I agree with Mason about the almost magical mixture of 66% hydrogen with 33% oxygen (Brown's Gas) that seems vastly superior to straight hydrogen inhalation machines. I have tried and extensively used three different hydrogen inhalation machines from three leading companies for three years. I started first on the Aquacure (Browns Gas), and then because my son needed it more than me, I switched to an expensive high output molecular hydrogen machine.

The reason I reached out to Masson was that I had just received a Hydrogen Technology machine but had not plugged it in yet, and I was wondering what to expect. Now ten days after the ultimate sleep therapy, breathing hydrogen and oxygen all night, I am in love with the machine. It already feels like it will add a decade to my lifespan.

I realized that my 6,000 dollars Vital Reaction hydrogen machine is inferior to what is available when one uses a good Brown's Gas machine.

Choosing between models that produce the same gas is like buying a car. If you have been driving a small car, which gets you everywhere, and suddenly you enter a Mercedes, you might not only reach your destination faster, but you will enjoy the journey with more ease and comfort.

I must say that anyone who has the financial muscle should go for the more expensive machine if one wants the ultimate healing machine by your bedside that is so whisper-quiet one can enjoy doing hydrogen therapy all night long as you sleep. With all the ridiculous things rich people spend money on these days, it is worth the extra investment.

However, I would take a 2,500 Brown's Gas machine (discounted 20% now) over a 6,000 dollars pure hydrogen machine any day of the week. In terms of medicine and health, a hydrogen medicine inhalation machine is the deal of a lifetime. A chance to extend our lives, a promise of a reduction in pain if one is suffering. And an excellent opportunity to heal if one is seriously ill. No gimmicks, no games. No unproven techniques. Biologically safe.

I finally have my long overdue miracle. Not having to wake up in pain has been a dream for many years.

When Browns Gas (BG) is used, the oxygen component would be a relatively low dose. George Wiseman, the AquaCure hydrogen inhalation device maker, recommends a gas flow of 18-20 L/h for the inhalation of BG for a normal-weight adult. Here there is a net increase of inhaled oxygen in the BG-air mixture increases from 21 to 23 vol%, as it is presumed that the BG contains 33 vol% oxygen (while in ambient air, the volumetric concentration amounts to 21 %).

The Hydrogen Technologies inhaler (seen above) is the Mercedes or BMW of the hydrogen inhalation machines. No other device comes close to touching it in terms of durability, quietness, and ability to run 24/7, so it is ideal for clinical environments.

The Molecular Hydrogen/Oxygen generator is made using the highest quality materials. Its patented generator design is made from stainless steel, which is used extensively throughout its devices. Their engineers believe that plastics and polycarbonates may negatively affect the gas's quality when used over extended periods.

The 810 Hydrogen Technology machine lists for 7,500 dollars. Mention my name, and you can receive a sizable discount. I have now been using it to do the ultimate sleep therapy for two months of nightly use, and I can tell you I am a new man. 99% of my pain is gone! However, the AquaCure offers the same gas and most of this machine's features for about a third the cost and produces the same gas.

The Hydrogen Tech generator is made using the highest quality materials. Its patented generator design is made from stainless steel, which is used extensively throughout its devices. Their engineers believe that plastics and polycarbonates may negatively affect the gas's quality when used over extended periods.

https://hydroproducts.info/

The Hydrogen Technologies machine produces up to 600 mL of hydrogen and 300 mL of oxygen per minute.

The large, 15 plate electrolysis cell and tank capacities in the Hydrogen Technologies machine allow for efficient gas production at low electrical demand and low operating temperature.

The AquaCure® (Model AC50) is an advanced, user-friendly HydrOxy for Health device. It was developed using over 30 years of Electrolyzer R&D experience and feedback from thousands of customers. See my site for a generous discount. http://hydroproducts.info

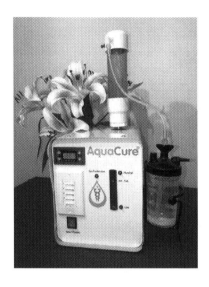

Retails for $2500, $2000 with a 20% discount through my site.

The AquaCure is a lower cost, practical, dependable, and versatile hydrogen inhaler that can give decades of trouble-free service and is designed to be the world's safest HydrOxy machine, with safety certification, pressure relief, pressure control, variable output, liquid level control, etc. It consumes about 200 watts of power to produce up to 833 mL (558 mL and 275 mL of oxygen) of HydrOxy gas per minute, making it one of the best options if you're looking to own a HydrOxy Inhalation machine.

It has 10 electrolysis cells inside a stainless-steel reservoir tank to efficiently minimize size. It has a commercial-grade white powder coated iron housing (not stainless steel like the Hydrogen Technologies)

because it is designed to have FULL safely and functionality at a lower cost.

While it is designed to keep the output gasses cool, the cooling fan is slightly louder than the Hydrogen Technologies. It has a lifetime warranty and a one-year satisfaction guarantee. It can be used continuously with minimal maintenance (just add pure water and rinse it every 100 hours).

The Hydrogen Technologies "Hydroqube" series (affectionately known as the "QB" series) is now available and is focused around the Proton Exchange Membrane (PEM) electrolysis technology of which we have put our style and engineering excellence standards on.

The machine produces a mixed atomic count of 66.6% hydrogen and 33.3% oxygen gas by splitting the distilled or demineralized water molecules. The retail price is 4,995 dollars. Mention my name for receiving a good discount.

Hydrogen inhalation goes perfectly with hydrogen water, otherwise known as hydrogen-rich water. For a few years, hydrogen water machines put low amounts of hydrogen into one's drinking water. Mine sits in the closet along with my water ionizer, which puts out high pH but low alkalinity and low hydrogen. Molecular hydrogen water machines cannot stand up to hydrogen inhalers or even high ppm hydrogen water tablets. The great advantage of the hydrogen-oxygen inhalers above is that they both also make your hydrogen oxygenated water. I only use hydrogen gas and am more than happy with the results.

If you need help implementing Hydrogen therapy, you can consult with me. I can help you decide which hydrogen inhaler is best for you, suggest the time spent breathing hydrogen gas to receive desired results, and what other therapies would support and fill out hydrogen therapy. See my consultation page on drsircus.com if you want my full support.

What to Expect from Hydrogen Inhalation Treatments.

- Detoxification.
- Restores youthfulness.
- Ulcers and sores healing.
- Firmer and thicker hair.
- Reduction in blood pressure.
- Slows down free radical damage.
- Lowers cholesterol levels.
- Help flush heavy metals from our bodies.
- Helps in the absorption of supplements.
- Improved allergies and asthma conditions.
- Better blood circulation.
- Lower saturated fat levels.
- Fewer body fatigues.
- Faster recovery from diseases.
- Improved peripheral circulation.
- Reduces cellulite and wrinkles.
- Improves memory in elderly.
- Boosts brain power.
- Reduces acidic condition.
- Improved constipation and diarrhea conditions.
- Improved blood glucose.

The more serious one's condition, the more one wants to combine hydrogen gas inhalation with hydrogen water. But when one's back is against the wall, magnesium bicarbonate water is much stronger therapeutically than hydrogen water.

Common testimonies using hydrogen include:

diminishing of numbness in extremities,
reduction and elimination of edema,
improvement of sinus problems,
more energy,
improvements in blood sugar,
reducing the need for insulin,
feeling the difference after the first treatments,
feelings of refreshment and lightness on one's feet,
improvements in skin conditions,
improvements in stamina,
feelings of body changes and energy levels, unstiffening
of knees and ankles,
stimulate youthful senses, more alert,
progress in circulation,
reduction of pain, headaches gone, decrease in need to
take painkillers,
skin appearance dramatically improved, healthier hair,
spots on face reduced,
reduction of constipation,
lesser feelings of depression,
neck no longer stiff and restoration of the full range of
movement, nails become stronger,
hair breakage and split ends minimized.

For inhalation, a 2-4% hydrogen gas mixture is typical because it is below the flammability level; however, some studies use 66.7% H2 and 33.3% O2, which is non-toxic but flammable if concentrated. This is not a problem because the gas coming out of machines is immediately diluted with room air. Even if one lights a match close to one's nose (not recommended) when using a hydrogen inhaler, it still would be rare to see a flame.

There are other methods to ingest or consume H2. Drinking hydrogen water using tablets that dissolve hydrogen into the water, injecting H2-dissolved saline (H2-saline), H2 baths, and dropping H2-saline into the eyes. Transdermal use of hydrogen and CO2 can be beneficial, as are hydrogen IVs. At this point, it is only being used in Japan and China.

Safety

If you look hard enough, you will find someone who says, 'We do breathe hydrogen: only trace amounts of hydrogen (H2) are present in the air, so we shouldn't breathe more.' Or crazy things like 'Breathing pure hydrogen will kill you,' as if you could breathe pure hydrogen from any of these inhalation machines, which you cannot. Pure oxygen would kill you; that is why there is always CO2 in oxygen cylinders.

Like in all devices, use caution. Read instruction books carefully before use. When using hydrogen-oxygen machines, extra caution is necessary, so follow instructions. Hydrogen and oxygen at a perfect mix (Stoichiometry) can explode, but that is not a problem with these inhalation machines. The mixture of gases gets diluted as one breathes in.

Hydrogen History.

13.8 billion years ago, the Universe was born, filling up physical space with hydrogen. Hydrogen is the most abundant element in the universe. All the hydrogen in the universe has its origin in the first few moments after the Big Bang. It is the third most abundant element on the Earth's surface after oxygen and silicon. Hydrogen is in all stars, and the planet Jupiter is composed mainly of hydrogen. Hydrogen and oxygen become water. Thus, hydrogen is the most abundant atom in our bodies.

In 1970, South African-born electrochemist John Bockris first used the term "hydrogen economy" in a speech. He later published a book describing what a solar-hydrogen-powered world might look like. But again, nothing changed.

An anti-inflammatory effect of hyperbaric hydrogen on a mouse model of schistosomiasis-associated chronic liver inflammation was also reported in 2001.[54]

In 2002, American economic and social theorist Jeremy Rifkin argued that hydrogen could take over from oil and that the future of energy lay in hydrogen-powered fuel cells.

The story of hydrogen as a medicine officially began in 2007 when Ohsawa and colleagues discovered that H2 has antioxidant properties that protect the brain against I/R injury and stroke by selectively neutralizing hydroxyl radicals.[55]

Hundreds of studies in the years since have been published showing hydrogen as a safe and effective medicine for over 150 disease models.

Today, many countries are getting officially into hydrogen for cars, trains, busses, and trucks.

Conclusion.

As the simplest and smallest element in the known universe, Hydrogen is going to lead to "a brighter, more intellectual, and healthier way of living," write Xuejun Sun, Shigeo Ohta, and Atsunori Nakao in their foundational book *Hydrogen Molecular Biology and Medicine*. They say this because hydrogen is a medicine, the safest one in existence. Our existence depends on hydrogen.

Inflammation, Free Radical Damage, Oxidative Stress, and Hydrogen.

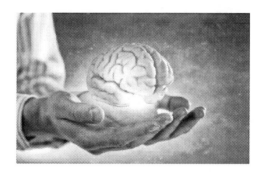

Since Dr. Denham Harman published his free-radical theory, more than 300,000 research studies confirm that free radicals are the leading cause of diseases and aging. Oxidative stress is a plague on modern people. Whether it is the toxic pollution in the air you breathe 24/7, medically induced radiation,[56] pharmaceutical medicines, chemotherapy, even WI-FI and other EMF pollution generating devices, you expose yourself to oxidative stress,[57]

Numerous health conditions, including chronic fatigue syndrome, fibromyalgia, diabetes, Alzheimer's disease, anxiety, insomnia, cancer, and just about every ailment, you can imagine roots in oxidative stress. It is directly or indirectly caused by anticancer drugs' toxicity in noncancerous tissues. During chemotherapy, oxidative stress-induced lipid peroxidation generates numerous electrophilic aldehydes attacking cellular targets.[58] Oxidative stress, induced by all prescribed medications, acts as a source of many dreadful diseases. Reactive metabolites formed during this process cause oxidative stress and impair drug-metabolizing enzymes' function, leading to toxicity.[59]

The World Health Organization calls pollution a more significant global threat than Ebola and HIV. According to its recent report, one in four deaths among children under five are from environmental hazards such as air pollution and contaminated water. Poisons in the air and water create oxidative stress, which leads to disease, cancer, and death. Epidemiological studies have shown a clear association between cardiovascular morbidity, decreased lung function, increased hospital admissions, mortality, and airborne concentrations of photochemical and particulate pollutants.

Heavy Metals

Heavy metals clog up receptor sites, break, and bend sulfur bonds in enzymes like insulin, damage the DNA, and in general, muck up everything and anything to do with healthy biological life.

Dr. Harold Buttram says, "Much of the current complacency about human chemical exposures is due to the method of toxicity testing, based on animal studies, in which a single chemical is evaluated to find an estimated 'safe' level for human exposure. The Delaney Amendment requires such testing of 1958, which demands testing potentially toxic chemicals for carcinogenic properties. There are several flaws and inadequacies in this system. it does not consider the effects of simultaneous exposures to multiple chemicals and their additive effects."

Most cancer patients have psychological and physiological stresses that contributed to cancer formation within the body. Psychological stresses include inescapable shock, repressed emotional pain and anger, depression, isolation, poor sleep, emotional trauma, and circumstantial life stresses. Physiological stresses include poor nutrition, chemicals, toxins, EMF radiation, parasites, liver, colon, kidney disease, and lack of exercise.

Cancer is a prime example of heavy-metal toxicity, free-radical damage, pathogen infection, mineral and vitamin deficiencies, inflammation, mitochondria dysfunction, immune system depression, genetic mutation, cell wall damage, and oxidative stress.

Cancer treatment can be approached in many ways. The best way is to simultaneously address all these problems, which a hydrogen-led protocol would do. Common environmental factors that contribute to cancer death include tobacco (25-30%), diet and obesity (30-35%), infections (15-20%), radiation (both ionizing and non-ionizing, up to 10%), and stress, lack of physical activity, and environmental pollutants and 5-10% due to genetics.

Studies have shown an increase in the general population's brain tumors over the last ten or more years. During the previous two decades, extensive research revealed how oxidative stress leads to chronic inflammation, leading to most chronic diseases, including cancer.

Cellular exposure to ionizing radiation leads to oxidizing events that alter macromolecules' molecular structures through direct radiation interactions targeting macromolecules or water radiolysis products. Further, the oxidative damage may spread from the targeted to neighboring, non-targeted bystander cells through redox-modulated intercellular communication mechanisms. People who started using cell phones at an earlier age have a greater chance of developing a brain tumor than people who started late (during their adult years).

Reactive oxygen species (ROS) are a by-product of normal metabolism. Even under pristine conditions, when our cells use glucose to make the energy, we create a cascade of free radicals that cause oxidative stress. The more sugar we consume, the greater our oxidative stress. When our immune system is fighting off bacteria and inflammation, we suffer from increased oxidative stress. When our bodies detoxify pesticides, herbicides, fungicides, we create oxidative stress.

Oxidation increases when we are physically and/or emotionally stressed. However, as long as we have enough antioxidants, a balance is maintained without damage. Oxidative stress happens when free radicals exceed antioxidants. Oxidative stress is an imbalance between free radicals' production and the body's ability to counteract or detoxify their harmful effects through neutralization by antioxidants.

Hydrogen the Ultimate Anti-Aging Medicine.

Hydrogen is a dream come true for the anti-aging crowd. We have the knowledge and the technology to extend our lives, enough medicinal firepower to press back against the hard edge of time.

The tons of money poured into developing life-extension treatments and technology will not replace health fundamentals. There is no single factor contributing to longevity; however, that does not stop us from knowing how to live longer and healthier. Some medical scientists believe that we may have reached our maximum limits for height, lifespan, and physical performance.

Hydrogen Medicine pushes against time. It is a wind that blows ICU patients away from death's door and reasonably healthy people back toward youth. Dr. Nick Delgado says, "Mounting evidence suggests hydrogen therapy may be the fountain of youth we have been searching for since the dawn of time. Hundreds of studies confirm it's not only safe, but it's also highly effective for the treatment of numerous diseases, for enhancing energy and sports performance, and for the promotion of optimal health and longevity."

Dr. Mercola said recently, "If there's ever a pill that will ensure extended youth, everyone will likely want it. The fact is, your lifestyle and the choices you make every day play an incredible role in how you will age, and I doubt a drug will ever be devised, allowing you to be a junk food-eating couch potato and still age in reverse. Of crucial importance is keeping your mitochondria healthy, and lifestyle strategies such as diet and exercise are key for this."

Oxygen is a drug, and hydrogen will also be one, especially in ICU, emergency, and surgery rooms. Hydrogen therapy is not snake oil. Hydrogen runs the universe, cars, trucks, rockets, and, if taken in sufficient quantities, will extend a person's life because of its protective effects and because it keeps the mitochondria healthy and humming along, precisely what Mercola suggests.

Scientists have a pipeline full of promising anti-aging compounds waiting for human trials, yet un-patentable hydrogen is not on their list. According to Joe Betts-Lacroix, the antiaging business is "an $8 billion industry of stuff that doesn't work." He believes the key to longevity is the prevention of degenerative disease.

Hydrogen works for prevention and healing, offering high medical horsepower even in ICU and emergencies. Hydrogen, the smallest, simplest atom and molecule in existence (H- and H2), is the safest anti-aging medicine for prolonging our existence.

Molecular hydrogen acts as a mighty antioxidant, helping defend cells and genes from the damage caused by harmful free radicals. Combined with its anti-inflammatory effects, this property helps enhance longevity because aging is caused by tissue degeneration, oxidative stress, and inflammation.

It was no surprise that in July 2016, a Japanese study concluded that hydrogen showed "substantial evidence," indicating vascular benefits. These medical scientists showed specific long-lasting antioxidant and anti-aging effects on vascular endothelial cells through the Nrf2 pathway, even after transient exposure to H2. They concluded that hydrogen-rich water is a functional drink that increases longevity.[60]

A study, "Effects of Hydrogenated Water on Intracellular Biomarkers," shows that hydrogenated water increases telomerase activity, an enzyme that promotes healthier telomeres - the caps at the end of chromosomes that impact aging. It concludes that molecular hydrogen helps protect telomeres from degradation, "contributing to age regulation improvement." These scientists also investigated hydrogen's effect on insulin signaling.

In terms of aging, hydrogen gas or water's antioxidant effect would diminish our telomeres' erosion. Telomeres in most human cells shorten with each round of DNA replication because they lack the enzyme telomerase. This is not, however, the only determinant of the loss of telomeric DNA. Oxidative damage is repaired less well in telomeric DNA than elsewhere in the chromosome. Oxidative stress accelerates telomere loss, whereas antioxidants decelerate it.[61]

For $8,000, Ambrosia will infuse older patients with a young donor's blood serum. You can treat yourself every day for much less money and include the entire family for no extra cost. Everyone can live longer with hydrogen.

Silicon Valley executives follow weird revitalization fads. They think the code of aging can be hacked, and death is made optional. Just wait till they discover the anti-aging miracle of molecular hydrogen, which makes surviving under great stress much more manageable. Just ask deep-sea divers what they are breathing to stay alive at depths of almost 2000 feet. Hydrogen!

Until now, we have made little progress in extending the outer limit of the human lifespan. Yes, more people live longer because we have gotten better at nutrition, curing acute conditions such as infections, and treating a handful of chronic diseases. But the maximum reported age has plateaued at around 115 years.

Now with hydrogen inhalation therapy, supported with increased oxygen delivery and increased CO_2 retention, we can stretch our years beyond all expectations while we enjoy peak health during that time.

The world is graying at a break-neck pace. Most healthcare dollars are spent on extraordinary care in the last two to three years of life— specifically on cancer and intensive care for heart and stroke patients. Over one hundred million Americans are currently being treated for one or another degenerative disease at a health care cost of more than $700 billion per year. Medical insurance costs are shooting up at a rate that means most people will not afford the cost of aging.

All we need is to do 10 minutes of large body movements every day to increase the blood flow, flush out the toxins, and inject the entire body with energy-giving oxygen and nutrients.

Miranda Esmonde White

The opposite of a tired, rundown aging cell is a metabolically active cell charged with vital mitochondria. The hallmarks of strong, metabolically active mitochondria are a myriad of factors, including:

> high levels of bicarbonate and carbon dioxide (alkaline conditions),

healthy oxygen transport,
high levels of magnesium ions,
healthy sleep and breathing patterns,
average body temperature
healthy levels of nutritional agents in general.

Disease states, including cancer, are attributable to mitochondrial dysfunction and oxygen deficiency (hypoxia), as eminent scientists like Albert Szent-Gyorgyi, Otto Warburg, Emmanuel Revici, and Linus Pauling assert. Nearly 80 years after Otto Warburg proposed cancer was caused by mitochondrial dysfunction, recent research from the Karolinska Institute in Sweden, Boston College, and Washington University School of Medicine have now thoroughly revived his theory.

Scientists think aging can be attributed to free radical damage and chronic low-level inflammation, which builds over time. All the unwelcome effects of aging, including metabolism slowdown, stiffness, frailty, aches, and pains, are caused by toxic build-ups and nutritional deficiencies, of the most necessary nutrients, like oxygen, carbon dioxide, hydrogen, and magnesium. The best methodology for reversing our aging process is to combine therapies that reinforce and multiply the other.

Living longer is all about You.

If you want to live to one hundred, you may want to watch more than your diet. A study published in the International Psychogeriatrics of people in remote Italian villages who lived past ninety found that these older folks still had a purpose in life, were still active, and had personality traits that included controlling and domineering stubbornness.

The oldest adults had other common qualities, including positivity, a strong work ethic, and close bonds with family, religion, and the countryside. Most of the older adults in the study were still active, regularly working in their homes or on their land. This gave them a purpose in life, wrote the study authors, even after they reached old age.

In my book *Heart Health*, I write that people with open and radiating hearts stay young forever. When wide open, the spiritual heart represents a fountain of youth and a force that helps us resist environmental insults, infections, and disease. There is nothing like love and an unselfish disposition to sustain a person through the decades.

Unconditional love is your immune system's most powerful stimulant.

Dr. Bernie Siegel

Dr. Norman Shealy and Dr. Caroline Myss believe that loving others and being loved are vital factors in improving the immune system, adding to life expectancy, and creating overall happiness. What does love have to do with stress-free living? "Everything!" says Dr. Brenda Schaeffer.

Feeling young and staying young is not a fool's dream. Folks who feel "young at heart" are more likely to live to a ripe old age. Seniors feeling three or more years younger than their actual age experienced a lower death rate over eight years than people who either felt their full age or a little older. About 25 percent of people who felt older than their actual age died, compared with about 14 percent of people who felt younger than their actual age and 19 percent who felt their age. More than two-thirds of participants felt three or more years younger than their actual age, while about a quarter felt their age. About 5 percent felt more than a year older than their actual age.

The findings show how powerful optimism can be when it comes to a person's overall health, said James Maddux, professor emeritus of psychology and senior scholar at the Center for the Advancement of Well-Being at George Mason University Fairfax, Va.

"Optimism in many ways is a self-fulfilling prophecy," he said. "If you feel your life and your health are largely under your control, and you believe you are capable of doing things like managing stress, eating right, and exercising, then you are more likely to do those things."

More than twice as many people who felt older than their actual age died from heart-related illness, compared to those who felt young — 10.2 percent, compared to 4.5 percent.

Anti-Aging Hydrogen and Oxygen Testimony.

"I started drinking hydrogen and oxygen (HydrOxy) water in 2005. Eventually, in March of 2016, I started breathing HydrOxy for several hours each day while working at the computer. After several months, I felt rejuvenated by one year for every month that I inhaled the gas."

"I just had a fascinating thing that happens to me. I was recently pounding down a steel post with an improvised post pounder and sprained my shoulder badly. I could not lift my arm without severe pain. Then something amazing happened. My shoulder sprain healed in three days! I've never had a sprain heal that fast!"

My eyesight has improved. I've worn glasses since nine years old. I don't wear glasses anymore except when driving.
My psoriasis is gone; no more thick white peeling skin on elbows, knees, and feet. This happened within three weeks of my starting to breathe HydrOxy.
My skin is smooth and supple, with age wrinkles gradually disappearing.
My scars (I have had since childhood) seem to be disappearing.
My 'age spots' are disappearing.
My neuropathies are gone. I'm grateful to feel my left hand and shins again.
My hair continues to darken (now salt and pepper instead of straight grey).
My hair seems to be growing back (thickening and growing on my bald top).

My tinnitus is still there but barely noticeable occasionally."
I'm losing fat and gaining muscle even without dedicated exercise.

He updated this in October 2017:
My warts are gone (hand warts and planter's wart).
My hair is definitely growing back.
My constipation is gone.
My arthritis is gone.
I still haven't been sick (not even a sniffle) since 2005 (NO drugs or flu shots).
I've lost 40 excess pounds (down to 180 from 220). My heart murmur is gone.

George Wiseman

Hydrogen, Bicarbonate, Magnesium, and ATP.

The Nobel Prize was awarded to Dr. Peter Mitchell in 1978 for his theory of chemiosmosis. According to his model, Hydrogen is essential in ATP production in the mitochondria, the energy source in the cells, and the body. This works through hydrogen dehydrogenase, a flavoprotein catalyzing the conversion of NAD+ to NADH by molecular hydrogen (H2), $H2 + NAD+ \rightarrow H+ + NADH$.

Our body depends upon the energy gained from the food we eat. Food undergoes many processes and then finally ends in the release of ATP. Food is a primary source of hydrogen. If it is fresh and uncooked, it provides an abundance of hydrogen. The hydrogen in food is tied up in complex molecules that must be metabolized (broken down) to release the hydrogen.

When your body burns hydrogen and oxygen, it generates the energy you need for the process of life. Water has both the fuel hydrogen and oxygen, which provides the fire of oxidation. The word hydrogen comes from the Greek, meaning "water-former." Water forms when hydrogen is burned (oxidized) by oxygen. It is created every day in our bodies as we burn hydrogen to create ATP. Hydrogen and oxygen participate in a continuous cycle that generates water and energy.

Magnesium researchers have long maintained that magnesium's roles (including calcium regulation) can only be fully utilized when the body is adequately hydrated. Water's importance cannot be understated for the pH balance, carrying toxic elements out of the body, and fully utilizing minerals ATP energy production.

An open-label trial of drinking 1.0 liter of hydrogen infused water (molecular hydrogen) per day over 12 weeks with fourteen patients, five having progressive muscular dystrophy, four with polymyositis dermatomyositis, and five having mitochondrial myopathies. Drinking hydrogen water improved mitochondrial function in mitochondrial myopathies and reduced the inflammatory processes in polymyositis and dermatomyositis.[62]

To understand the concept and practice of increasing cellular lifespans, health, and vitality, one must look deep into the cell's mitochondria, the Krebs cycle, the electron transport system, and how hydrogen plays a crucial role in creating ATP.

Negative hydrogen ions can generate ATP in the mitochondria and trap free radicals effectively. The H- changes NAD+ in the mitochondria to NADH. The NADH has then been processed in the electron transport system to produce 3 ATP molecules. Typically the Krebs cycle produces NADH, but the H- makes it possible to bypass the Krebs cycle by recycling NAD+ to NADH.

Damaged mitochondria provoke pathogenesis seemingly unrelated disorders such as:
    schizophrenia,
    bipolar disease,
    dementia,
    Alzheimer's disease,
    epilepsy,
    migraine headaches,

strokes,
neuropathic pain,
Parkinson's disease,
ataxia,
transient ischemic attack,
cardiomyopathy,
coronary artery disease,
chronic fatigue síndrome,
fibromyalgia,
retinitis pigmentosa,
diabetes,
hepatitis C,
primary biliary cirrhosis.

Dr. Michael R. Eades says, "high-energy electrons passed down the inner mitochondrial membrane occasionally break free. When they break free, they become free radicals. These rogue free radicals can then attack other molecules and damage them. Because these free radicals are loosened within the mitochondria, the closest molecules for them to attack are the fats in the mitochondrial membranes. If enough of these fats are damaged, the membrane ceases to work properly. Then the entire mitochondrion is compromised and ceases functioning. If enough mitochondria die, the cell doesn't work and undergoes apoptosis, a kind of cellular suicide. This chronic damage and loss of cells is the basic definition of aging."

Hydrogen sulfide and mitochondrial function.

Hydrogen sulfide (H_2S), at low concentrations, acts as an electron donor to the mitochondria and serves as an inorganic substrate for mitochondrial electron transport. H_2S donation can exert therapeutic effects. The idea that H2S can be used to induce on-demand metabolic suppression ("hibernation") has been proposed in other conditions. In yet other conditions (e.g., inflammation, septic shock, burn injury, and certain forms of cancer), H_2S overproduction occurs, and H_2S biosynthesis inhibitors may be a future therapeutic approach. Hydrogen sulfide may play a wide-ranging role in staving off aging. However, it is much easier and safer to use molecular hydrogen.

Magnesium Bicarbonate - Medical Breakthrough

Increased oxidative stress, which correlates exponentially with pH changes into the acidic, is hazardous to the mitochondria, resulting in oxidative duress. When our tissues become too acidic and lack the magnesium necessary for ATP production, cellular metabolism drops off, leading to obesity and diabetes.

"Mg2+ is critical for all of the cells' energetics because it is required that Mg2+ be bound by ATP, the central high energy compound of the body. ATP without Mg2+ bound cannot create the energy normally used by specific body enzymes to make protein, DNA, RNA, transport sodium, potassium, or calcium in and out of cells. ATP without enough Mg2+ is non-functional and leads to cell death," writes Dr. Boyd Haley.

Magnesium bicarbonate is the Holy Grail of magnesium administration, offering a straight line into the cells and firing up one's mitochondrial energy factories. My highest recommendations are to make and drink hydrogenated and oxygenated water or magnesium bicarbonate water. Drink your way back to life. If one is doing water fasting, do it with these types of water.

Magnesium bicarbonate is a complex hydrated salt that exists only in water under specific conditions. The magnesium ion is Mg2+, and the bicarbonate ion is HCO3-. So, magnesium bicarbonate has two bicarbonate ions: Mg (HCO3)2.

Magnesium is the lamp of life and is a cofactor for more than 300 enzymatic reactions. Magnesium is crucial for adenosine triphosphate (ATP) metabolism, DNA and RNA synthesis, reproduction, and protein synthesis.

Magnesium and bicarbonate together work to combat the drop-in energy within the mitochondria during constant bombardment from toxins. It neutralizes the acid produced from metabolic processes and ATP hydrolysis, allowing more ATP to be hydrolyzed or more energy to be made. Magnesium bicarbonate buffers the mitochondria in body cells from excess acid concentrations, improving mitochondrial function and increasing ATP.

The ideal alkaline water is rich in magnesium and bicarbonate. Alkaline water machines that produce high pH water cannot hold a torch to waters high in magnesium and bicarbonate. The work of Dr. Russell Beckett, a veterinarian with a Ph.D. in biochemical pathology, paved the way to understand the significance of bicarbonate in conjunction with magnesium. Unique Water from Australia and Noah's Water in California, Donat Mg from Europe (natural spring waters), and a convenient magnesium bicarbonate concentrate made in Florida offer water that is medicine.

Magnesium and bicarbonate-rich mineral waters are easily absorbed. Because bicarbonates control your body's pH, sodium bicarbonate can rapidly alkalize your body far more effectively than a diet. I recommend a product called pH Adjust because, in addition to sodium bicarbonate, it has potassium bicarbonate (reducing the sodium load and supplying essential potassium) as well as a dose of magnesium. Magnesium bicarbonate can be used long-term.

Magnesium bicarbonate is one of the most convenient ways to replenish our bodies with the necessary magnesium: very high bioavailability, no calcium present, and bicarbonate help maintain blood pH. Magnesium bicarbonate is suited to help the body excrete an excess of calcium. The absolute best magnesium supplement is magnesium bicarbonate water, either in natural spring waters or in the water you make at home with the Magbicarb concentrate.

Bicarbonate.

When the body has sufficient bicarbonate, it resists better the toxicity of chemical insults. That is why the army suggests its use to protect the kidneys from radiological contamination.

Research published recently in the Clinical Journal of the American Society of Nephrology found having balanced bicarbonate levels in your body reduces the chances of early death. The study examined data compiled in the health, aging, and body composition study for 2,287 participants.

Dr. Kalani Raphael, professor at the University of Utah, and colleagues investigated pH, carbon dioxide, and bicarbonate in association with long-term survival. "Critically ill patients with severe acid-base abnormalities have a very low likelihood of surviving their illness."

Dr. Raphael found that low bicarbonate levels are linked to an increased risk for premature death by 24 percent.

Bicarbonate acts as a transporter of magnesium into the mitochondria. Magnesium does not readily reach the mitochondrion, but the magnesium will flood if plenty of bicarbonate is available. According to the Dietary Reference Intakes guide from the Institute of Medicine, magnesium influx is linked with bicarbonate transport. Bicarbonate acts to stimulate the ATPase by working directly on it.

The bicarbonate buffer system occurs in both intra- and extracellular fluids. It consists of carbonic acid (H_2CO_3) and sodium bicarbonate ($NaHCO_3$). If a strong acid is present, it reacts with sodium bicarbonate to produce carbonic acid and sodium chloride, minimizing the increasing concentration of positive hydrogen ions. If a sturdy base is present, it reacts with carbonic acid, producing sodium bicarbonate and water, minimizing the alkaline shift. Bicarbonate ion concentrations decrease the formation of acid by carbonic anhydrase enzyme (Le Chatelier's principle). In the presence of magnesium and bicarbonate ions, less acid is produced by the carbonic anhydrase enzyme.

H2O2 – Hydrogen Peroxide.

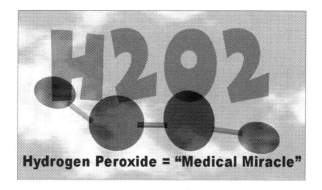

It seems hydrogen in all its forms (except H+ ions) is beneficial for our health as a potent medicine, even against cancer. Hydrogen peroxide (H2O2) ranks up as one of the best household remedies and is one of the few "miracle substances" still available to the public; it is safe, readily available, and dirt cheap. And best of all, it works! Did you know that you had your first sip of hydrogen peroxide shortly after you took your first breath? That is right, mother's milk (especially colostrum) contains high concentrations of H2O2."

Hydrogen Peroxide is the oldest oxygen therapy known. It provides the body extra needed oxygen. Oxygen, the most prescribed natural drug in hospitals, provides our primary source of energy. All the cells in our bodies need an adequate amount of this precious gas to survive and function normally. Oxygen performs medical miracles when it comes to healing. It means life, and H2O2 provides a low-cost method of getting more energy, more oxygen.

One does not have to be afraid of using H2O2 because it is a natural substance that the body makes to protect itself from hostile pathogens. It stimulates the immune system and natural killer (NK) cells, which attack cancer cells, preventing the spread throughout the body. The cells responsible for fighting infection and foreign invaders in the body (our white blood and T-cells) make hydrogen peroxide and use it to oxidize any offending culprits. The intense bubbling you see when hydrogen peroxide encounters a bacteria-laden cut or wound is the oxygen being released and bacteria destroyed. H2O2 is not an undesirable by-product or toxin but an essential requirement for good health.

Blood platelets release hydrogen peroxide on encountering particulates in blood. In the large intestine, acidophilus lactobacillus produces H2O2, keeping the ubiquitous candida yeast from multiplying. When candida escapes from the intestine, it leaves the natural control system and gains a foothold in the organs, causing chronic fatigue syndrome.

H2O2, whether coming from immune response or taken orally, intravenously, or through application to the skin, kills viruses and other pathogens via oxidation. At the same time, oxygen is boosted and revitalizes normal cells.

Supporters consider H2O2 as one of the most remarkable healing miracles of all time. It belongs right next to sodium bicarbonate, potassium bicarbonate, iodine, selenium, sulfur, and magnesium as necessary medicines for every medicinal cabinet. These are all superhero medicines because they feed the roots of life.

Like sodium bicarbonate and magnesium, hydrogen peroxide has many potential applications with little to no risks. The FDA (not my favorite organization) has given hydrogen peroxide the GRASS designation, which means recognized as safe. One of the advantages of peroxide therapy is that it is incredibly accessible like bicarbonate is.

Hydrogen peroxide's chemical formula is H2O2. It contains one more atom of oxygen than does water (H20). Hydrogen peroxide is odorless and colorless but not tasteless. When stored under the proper conditions, it is a very stable compound. When kept in the absence of light and contaminants, it breaks down very slowly at the rate of about 5 to 10% a year.

When exposed to other compounds, hydrogen peroxide dismutase's readily. The extra oxygen atom is released, leaving H20 (water). In nature, oxygen (02) consists of two atoms--a very stable combination. However, a single atom of oxygen is very reactive and called a free radical. One excellent way of taking peroxide is in a bath (up to a pint of 35% in a full bath). Better jet 1kg of sea salt + 500gr bicarbonate +1 pint 35% H2O2.

Hydrogen peroxide is a reactive chemical that is toxic when concentrated, yet we purchase H2O2 in a concentrated form of 35 percent. Thus, only diluted amounts of H202 (3 %) can and should be ever introduced into the body, meaning one must mix one ounce of 35 percent food grade peroxide with 11 ounces of pure water. 35 % food grade is the only grade used for internal use after being reduced to 5 %. 35 percent will burn the skin and is flammable, so please be careful to dilute properly.

The English medical journal, Lancet, reported in 1920 intravenous infusion of H2O2 to treat pneumonia the epidemic following World War I. In the 1940's Father Richard Wilhelm, the pioneer in promoting peroxide, treated everything from bacterial-related mental illness to skin disease and polio. However, much of the interest in hydrogen peroxide waned in the 1940s when prescription medications came on the scene.

In the last 25 years, thousands of articles related to hydrogen peroxide appeared in the standard medical journals. Thousands more involving its therapeutic use are printed in alternative health publications.

H202 works wonders on a multitude of health problems. It does so by increasing tissue oxygen levels. After reading the last chapter, it is evident that our oxygen needs aren't met. Hydrogen peroxide is one of many components that help regulate the amount of oxygen getting to our cells. Its presence is vital for many functions, such as producing thyroid hormone and sexual hormones and dilating blood vessels in the heart and brain. It improves glucose utilization in people with diabetes.

Note: Don't use chlorinated tap water to dilute the peroxide!

When using peroxide as a stand-alone healing system (see the One-Minute Cure by Madison Cavanaugh), the main recommendation is to start three drops three times a day and day by day work up to 25 drops three times a day. Then one backs down to the three drops a day and either continues or stops.

Some cannot reach high dosages because of nausea. When using other oxygen-gaining methods, like taking bicarbonate, practicing slow breathing, and increasing iodine and magnesium (which also increase oxygen), it will not be necessary to use so much peroxide.

An alternative method is to use H_2O_2 is using a spray/nasal bottle full of 3% solution and spray into the mouth 5-10 times a day and breathe the mist deep into the lungs.

Thomas Levy, M. D, a board-certified cardiologist, believes, as do others, that H_2O_2 (hydrogen peroxide) is another at-home treatment that can wipe out any virus, including the coronavirus. The key ingredient in this treatment is common household 3% hydrogen peroxide, and this is the same substance that can be purchased in a 32-ounce plastic bottle at Walmart for 88 cents or at Walgreens for under $1.00.

Because hydrogen peroxide consists of a water molecule (H_2O) with an extra oxygen atom (H_2O_2), this extra oxygen atom is what makes it so deadly for viruses. The great news is that there is a safe and simple way to administer this common substance, almost as effective as an IV. Like bicarbonate and glutathione, H_2O_2 can be nebulized.

Cancer, Inflammation, and Hydrogen.

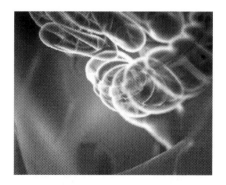

Numerous publications reveal the biological and medical benefits of H2 reduce oxidative stress by direct reactions with strong oxidants and indirectly by regulating various gene expressions. Moreover, H2 functions as an anti-inflammatory and anti-apoptotic and stimulates energy metabolism.[63]

Because we can define cancer as inflammation, we can employ hydrogen to win our battle against it. Dr. Johannes Fibiger was a Danish scientist, physician, and professor of pathological anatomy who won the Nobel Prize in Physiology and Medicine in 1926 for achieving the first controlled induction of cancer in laboratory animals, a development of profound importance to cancer research.

In 1907, while dissecting rats infected with tuberculosis, he found tumors in three animals' stomachs. After intensive research, he concluded that the tumors, apparently malignant, followed an inflammation of stomach tissue caused by a worm's larvae, now known as *Gongylonema neoplasticum*. The worms had infected cockroaches eaten by the rats.

By 1913 he consistently induced gastric tumors in mice and rats by feeding them cockroaches infected with the worm. By showing that the tumors underwent metastasis, he added support to the then-prevailing concept that cancer is caused by tissue irritation.

Fibiger's work immediately led the Japanese pathologist Yamagiwa Katsusaburo to produce cancer in laboratory animals by painting their skins with coal-tar derivatives, a procedure Fibiger soon adopted himself. While later research revealed that the Gongylonema larvae were not solely responsible for the inflammation, Fibiger's findings were a necessary prelude to producing chemical carcinogens (cancer-causing agents), a vital step in developing modern cancer research.

In Latin, the word "inflammation" means "ignite, set alight," and like gasoline, that is what it does to cancer. A microenvironment of chronic inflammation sets the stage for cancer. Most importantly, inflammation promotes the spreading and mutating of cancer cells while pushing the mutations within the cancer cells' development.

Oxidative stress plays an essential role in the pathogenesis of inflammation. Since we have already found that hydrogen gas and hydrogen water eliminate free radicals and oxidative stress, we know it reduces inflammation.

According to the National Cancer Institute, considerable laboratory evidence from chemical, cell culture, and animal studies indicates that antioxidants may slow or prevent cancer development. Antioxidants are nutrients (vitamins and minerals) and enzymes (proteins in your body that assist in chemical reactions). They play a role in preventing the development of chronic diseases such as cancer, heart disease, stroke, Alzheimer's disease, Rheumatoid arthritis, and cataracts.

Hydrogen is the ultimate antioxidant, so it helps us win our war on cancer. Cancer patients interested in humane treatments and who do not want to suffer toxic chemotherapy and radiation will be highly interested in hydrogen cancer treatments. Backed by oxygen and CO2 therapy, a list of vital minerals, and even a concentrated form of medical marijuana as a natural form of chemotherapy, you will feel much better.

Cancer Starts with Inflammation.

There is a strong association between chronic, ongoing inflammation in the body and cancer occurrence. Biologists have followed the inflammation link down to the level of individual signaling molecules, providing evidence of carcinogenesis. We already know that inflammation is the root of the pain and most illnesses like diabetes and heart disease. Still, we are just beginning to pay attention to the significant and centralized role of inflammation in cancer development and sustainment.

According to Dr. Alexander Hoffmann, an assistant professor of chemistry and biochemistry at U.C. San Diego, "We have identified a basic cellular mechanism that may be linking chronic inflammation and cancer. Studies with animals have shown that a little inflammation is necessary for the normal development of the immune system and other organ systems," explains Hoffmann. "We discovered that the protein p100 provides the cell with a way in which inflammation can influence development. But there can be too much of a good thing. In the case of chronic inflammation, the presence of too much p100 may over-activate the developmental pathway, resulting in cancer."

Inflammation is associated with the development of cancer. Scientific American says, "Understanding chronic inflammation, which contributes to heart disease, Alzheimer's and a variety of other ailments, maybe a key to unlocking the mysteries of cancer." Inflammation is the fuel that feeds cancer. It certainly is a critical event in cancer development.

"Inflammatory responses play decisive roles at different stages of tumor development, including initiation, promotion, malignant conversion, invasion, and metastasis. Inflammation also affects immune surveillance and responses to therapy. Immune cells that infiltrate tumors engage in extensive and dynamic crosstalk with cancer cells," says researchers from Departments of Pharmacology and Pathology, School of Medicine, University of California in San Diego.

Dr. Sergei I. Grivennikov writes, "The presence of leukocytes within tumors, observed in the 19th century by Rudolf Virchow, provided the first indication of a possible link between inflammation and cancer. Yet, it is only during the last decade that unmistakable evidence has been obtained that inflammation plays a critical role in tumorigenesis. Some underlying molecular mechanisms have been elucidated. A role for inflammation in tumorigenesis is now generally accepted, and it has become evident that an inflammatory microenvironment is an essential component of all tumors. Only a minority of all cancers are caused by germline mutations, whereas the vast majority (90%) are linked to somatic mutations and environmental factors."

An inflammatory microenvironment inhabiting various inflammatory cells and a network of signaling molecules are also indispensable for the malignant progression of transformed cells, attributed to the mutagenic predisposition of persistent infection-fighting agents at chronic inflammation sites. Chronic inflammation is a slow, silent disturbance that never shuts off. Often a patient cannot feel it. Usually, you cannot be tested for it.

Research regarding inflammation-associated cancer development has focused on cytokines and chemokines and their downstream targets in linking inflammation and cancer. Chronic inflammation due to infection or conditions like chronic inflammatory bowel disease is associated with 25 percent of all cancers. Researchers at the Ohio State University Comprehensive Cancer Center found that inflammation stimulates a rise in levels of a molecule called microRNA-155 (miR-155). In turn, it causes a drop in levels of proteins involved in DNA repair, resulting in a higher rate of spontaneous gene mutations, which can lead to cancer. "Our study shows that miR-155 is dysregulated by inflammatory stimuli and that over-expression of miR-155 increases the spontaneous mutation rate, which can contribute to tumorigenesis," says first author and post-doctoral researcher Dr. Esmerina Tili. "People have suspected that inflammation plays an important role in cancer, and our study presents a molecular mechanism explaining how it happens."

Dr. Vijay Nair's book *Prevent Cancer, Strokes, Heart Attacks, and Other Deadly Killers* says, "Colon cancer, stomach cancer, esophageal cancer, lung cancer, liver cancer, breast cancer, cervical cancer, ovarian cancer, prostate cancer, and pancreatic cancer have all been linked to inflammation. This is great news because it means that cancer does not just strike out of nowhere. It's preventable!"

All types of inflammation can cause cancer.

Lung cancer caused by chronic smoke-induced inflammation.
Esophageal cancer caused by acid reflux-induced inflammation.
Stomach cancer caused by H. pylori (the bacterium that causes ulcers) -induced inflammation.
Bladder cancer caused by urinary tract infection-induced inflammation.
Liver cancer caused by hepatitis B or C-induced inflammation.
Lymphoma caused by Epstein Barr (a virus causing mononucleosis) induced inflammation.
Cervical cancer caused by Human papillomavirus (the virus that causes genital warts) -induced inflammation.
Kidney cancer caused by kidney stone-induced inflammation.
Colon cancer caused by irritable bowel syndrome-induced inflammation.

Whether inflammation is caused by infection (such as hepatitis), a mechanical irritant (such as kidney stones), or a chemical irritant (such as stomach acid), the result is the same. Chronic, low-grade inflammation dramatically increases your risk of developing cancer."

Dr. Otis Brawley, chief medical officer of the American Cancer Society, said he believes aspirin's anti-inflammatory properties may help prevent heart disease and cancer. "Inflammation may not cause cancer, but it may promote cancer—it may be the fertilizer that makes it grow," Dr. Brawley said.

A new MIT study[64] offers a comprehensive look at chemical and genetic changes that occur as inflammation progresses to cancer. One of the most significant risk factors for liver, colon, or stomach cancer is a chronic inflammation of those organs, often caused by viral or bacterial infections. Orthodox cancer treatments do not treat inflammation; thus, they do not really treat cancer.

The role of heavy metals is significant in the rise of cancer rates.[65],[66] We are poisoning the world repeatedly with heavy metals, and our brain cells and other tissues are suffering from it. Over 80% of heavy metals are removed from the body via the friendly bacteria in the gut. Still, unfortunately, we have maniacs controlling western medicine encouraging doctors to overuse antibiotics, which kill off the gut's friendly bacteria. Heavy metal contamination creates inflammation!

2% H2 treatment effects were investigated on survival rate and organ damage in the zymosan-induced generalized inflammation model. Zymosan, prepared from baker's yeast (Saccharomyces cerevisiae), is a reagent widely used for many years in inflammation and immunology research.

The beneficial effects of H2 treatment in zymosan-induced organ damage were associated with decreased levels of oxidative product, increased activities of the antioxidant enzyme, and reduced levels of early and late pro-inflammatory cytokines in serum and tissues. H2 treatment protected against multiple organ damage in a zymosan-induced generalized inflammation model, suggesting the potential use of H2 as a therapeutic agent in the therapy with inflammation-related multiple organ dysfunction syndromes.[67]

Hydrogen for Psychiatric Disorders.

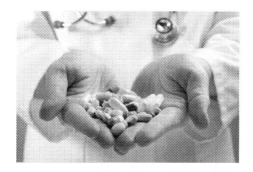

Antidepressants are among the most commonly prescribed drugs in the United States, but their side effects and trial-and-error nature often leave something to be desired. According to some studies, they are only about 50 percent more effective than placebo. Until recently, most doctors and patients have still thought that these drugs are the best treatment for a crippling disease. Now hydrogen gas promises to replace these obscene drugs.

One in six Americans overall regularly takes some type of medication in this category. However, children are now being swept up in pharmaceutical terrorism. The mental health watchdog group Citizens Commission on Human Rights is drawing attention to the fact that more than a million kids younger than six in our nation are currently taking these psychiatric drugs. Data from IMS Health shows that the drug situation worsens as kids get older, with 4,130,340 kids aged 6 to 12 taking some psychiatric drug type. This number is alarming, considering the horrifying side effects and ineffectiveness of these harmful drugs.

Babies Given Psychiatric Drugs.

Around half of these children are four to five years old, and an incredible 274,804 are younger than a year old. The number rises for toddlers aged two to three, with 370,778 kids in this category taking psychiatric drugs overall. The most prominent type of psychotic drug given to children is anti-anxiety drugs. Just over 227,132 babies under one year old and 248,000 of those aged four to five take these medications. However, experts believe these estimates are low. The actual numbers are much higher, due partly to the tendency for some doctors to hand out psychiatric medications for "off-label" uses.

In Europe, 4 out of 15 people suffer from major depression and anxiety. Neuropsychiatric disorders are the second cause of disability in Europe and account for 19% of cardiovascular diseases with only 4%. In 28 EU countries with 466 million people, at least twenty-one million were affected by depression, almost 80% are men. The treatment of psychiatric disorders is costly. Europe's total annual cost of depression was around 118 billion Euros in 2004, corresponding to 253 Euro per inhabitant. The price of depression corresponds to 1% of the entire economy of Europe.[68]

Hydrogen for the Brain.

Hydrogen gas, because of its anti-inflammatory and anti-oxidative properties, is helpful for neurological and psychiatric disorders. Data is emerging on the presence and impact of oxidative stress among various psychiatric disorders, including bipolar disorder, schizophrenia, and autism. Therefore, it is hypothesized that hydrogen molecule administration may have potential as a novel therapy for bipolar disorder, schizophrenia, and other concurrent disorders characterized by oxidative, inflammatory, and apoptotic dysregulation.[69]

Because of its low molecular weight, hydrogen can quickly diffuse across the blood-brain barrier, which allows it protects cells against degeneration and improves brain function. Chen et al.[70] found the protective effect of hydrogen in the brain is accompanied by reducing oxidative stress and blood glucose levels after dextrose injection in rats. Drinking hydrogen-rich pure water prevents superoxide formation in brain slices of vitamin C-depleted SMP30/GNL-knockout mice during hypoxia-re-oxygenation[71]. Molecular hydrogen has also prevented cognitive decline. Consumption of hydrogen water suppressed the increase in oxidative stress and prevented stress-induced impairments in hippocampus-dependent learning tasks during chronic physical restraint in mice[72].

Specifically, both bipolar disorder and schizophrenia are associated with increased oxidative and inflammatory stress. Moreover, lithium, commonly administered for treating bipolar disorder, affects oxidative stress and apoptotic pathways, as do valproate and some atypical antipsychotics for treating schizophrenia. Molecular hydrogen has been studied pre-clinically in animal models to treat some medical conditions, including hypoxia and neurodegenerative disorders. There are intriguing clinical findings in neurological disorders, including Parkinson's disease.

Mitochondrial dysfunction in schizophrenia is frequently reported. Moreover, mitochondrial disorders can present with psychosis. mtDNA plays a role in the neurobiology of schizophrenia. Mitochondrial gene expression is changed in schizophrenia. The number of mitochondria in schizophrenia is reduced compared to normal controls. As we have seen, hydrogen has a positive effect on the mitochondria to be utilized as the first line of schizophrenia treatment.

In Parkinson's disease, increased oxidative stress indexed by elevated lipid peroxidation and decreased glutathione levels in the substantia nigra are part of PD's known pathogenesis. Hydrogen water also prevents a rat model of Parkinson's disease[73] and increases survival after cerebral ischemia/reperfusion[74]. It down-regulates the marker of oxidative stress in dopaminergic neurons within the substantia nigra of animal models of Parkinson's disease.

Conclusion.

Everyone seeks health and vibrant life. Chronic oxidative stress and inflammation cause deteriorations in the central nervous system function, leading to a low quality of life (QOL). In healthy individuals, aging, job stress, and cognitive load over several hours also increase oxidative stress, suggesting that preventing the accumulation of oxidative stress contributes to ameliorate QOL and the effects of aging.

One study investigated the effects of drinking hydrogen-rich water (HRW) on the QOL of adult volunteers using psychophysiological tests. In this double-blinded, placebo-controlled study, results suggest that HRW may reinforce QOL through effects that increase central nervous system functions involving mood, anxiety, and autonomic nerve function.[75]

Less Food and Less Diabetes with Hydrogen.

Hydrogen is the most common element in the human body, being the majority stake or component of water, DNA, and most other organic molecules. Macro-nutrients, such as carbohydrates, proteins, and fats, contain hydrogen as part of the chemical structure. This means that every food contains some hydrogen.

Fruits and whole grains are healthy sources of carbohydrates, which in turn, are sources of hydrogen. Carbohydrates, also called sugars, are molecules that contain carbon, hydrogen, and oxygen. The chemical structure for glucose, the simple sugar your cells use to metabolize energy, has six atoms of carbon, twelve atoms of hydrogen, and six atoms of oxygen.

Meat, poultry, fish, dairy, and legumes are proteins made up of amino acids that are sources of hydrogen. The chemical structure of amino acids may differ, but all amino acids contain carbon, hydrogen, oxygen, and nitrogen. Fats, also called lipids, are macro-nutrients made up of carbon, hydrogen, and oxygen.

Hydrogen's role in the food chain receives little attention, yet it is an essential ingredient in our food, as Patrick Flanagan, the original hydrogen researcher, states. When we look at the dynamics of a healthy person inhaling and drinking hydrogen, we see that providing hydrogen in a gas form substitutes, to an extent, the hydrogen the body must strip off of foods during digestion.

People should rightfully expect to be nurtured and sustained by hydrogen. A know sixty-year-old man digging out a foundation (with lots of young helpers) could keep up with the best of them being at it all day long. This old young man was eating only one meal a day but spent two hours inhaling hydrogen and oxygen and drinking hydrogen water.

This story is significant in life extension because less is more, that restricting calorie intake increases the lifespan in every species studied. The rough rule of thumb restricts calorie intake by 30% and sees up to a 30% increase in lifespan. The more we run on hydrogen, which comes from sources that do not require insulin, the more heightened hydrogen's anti-aging effect. Researchers discovered that NAD+ levels decline with age but are raised by calorie restriction and exercise.

Insulin's job is to mobilize the body to respond to food intake. Like a warehouse overseer receiving a stock delivery, the hormone released into the blood ensures many systems mobilize quickly. The insulin 'receptor' conveys these signals to the body tissues, so nutrients are used as needed or stored as fat.

Calorie restriction induces a reduction in the insulin signaling pathways (both through IGF-I and insulin). This reduction in insulin signaling is one of the primary mechanisms through which calorie restriction increases lifespan.

High blood insulin levels are usually the result of high blood glucose levels. High insulin levels signal a high "nutritional state," which shifts the cell's internal states toward increased growth levels and decreased repair levels. Increases in growth and decreases in repair equal increases in waste products that accumulate within a cell, causing aging at the cellular level.

Hydrogen has no calories yet can send rockets into space. Hydrogen burns cleaner than any other fuel. When we look at the fact that food is our fuel and hydrogen is the easiest and cleanest fuel to run on, we start to understand why hydrogen is such a functional medicine and why it will extend our lives.

Diabetes.

Talking about calorie restriction and its powerful positive effects on our bodies makes it the right moment to talk about diabetes and what many holistic practitioners have known for years. In Newcastle and Glasgow, England, doctors report a new study finding that nearly half of patients have reversed type 2 diabetes in a "watershed" trial.

In the study, people spent up to five months on a low-calorie diet of soups and shakes to trigger massive weight loss. Isobel for example, aged 65, had weighed 95 kg, lost 25kg, and no longer needed diabetes pills. She says: "I've got my life back." Isobel was one of 298 people studied. Her blood sugar levels were too high. Every time she went to the doctors, they increased her medication. So, she went on to the all-liquid diet for 17 weeks - giving up cooking and shopping, having four liquid meals a day.

The trial results, simultaneously published in The Lancet Medical Journal and presented at the International Diabetes Federation, showed:

46% of patients who started the trial were in remission a year later.

86% who lost 15kg (2ˢᵗ 5lb) or more put their type 2 diabetes into remission.

Only 4% went into remission with the best treatments currently used.

From Newcastle University, Prof Roy Taylor told the BBC, "Before we started this line of work, doctors and specialists regarded type 2 as irreversible."

There is evidence that molecular hydrogen makes the process of recovering from diabetes easier. Hydrogen improves obesity and diabetes by inducing hepatic FGF21 and stimulating energy metabolism in animal studies.[76] Supplementation of hydrogen-rich water improves lipid and glucose metabolism in patients with type 2 diabetes or impaired glucose tolerance.[77]

Hydrogen-rich water:

- reduces oxidative stress in the liver and decreases liver fat.
- suppresses body-weight gain and reduced blood glucose and triglyceride levels.
- shows a similar effect to diet restriction.
- increases the level of FGF21 (a protein that regulates energy expenditure, protects from obesity caused by overeating, and lowers blood glucose and triglyceride levels) in the liver.
- stimulates energy metabolism.

Scientists had struggled to understand how ingesting a small amount of hydrogen in hydrogen-rich water could have such significant clinical effects. One study showed that hydrogen is accumulated and reserved in the liver by attaching to glycogen molecules.

Organs and muscles in the body use glucose as fuel. Carbohydrates, protein, and fats need to be converted to glucose to be used. Glycogen is glucose stored in both muscles and the liver. From the liver, glycogen is released as energy to organs and muscles when required. H2 accumulates in liver glycogen stores after being consumed via Hydrogen Rich Water.

H2 is available to be used as an antioxidant and anti-inflammatory when needed for the body.

When any part of the body needs glucose, that organ or muscle is also receiving H2. Therefore, the body's antioxidant mechanisms are improved, decreasing free radical damage. As free radical damage contributes to the development of diabetes, the increase in antioxidant functions impedes the development of diabetes. Mice receiving H2 water had lower levels of molecules that indicate free radical damage; the H2 treated mice had levels of these molecules almost the same as non-diabetic mice.

Subcutaneous injection of H2 was also used in animal studies and "significantly improved type 2 diabetes mellitus and diabetic nephropathy-related outcomes in a mouse model. The bodyweight of H2 treated mice did not change throughout the experiment. Compared with the untreated control animals, glucose, insulin, low-density lipoprotein, and triglyceride levels in the serum were significantly lower in treated mice, whereas high-density lipoprotein cholesterol in the serum was significantly higher. Glucose tolerance and insulin sensitivity were both improved in H2 -treated mice. Diabetic nephropathy analysis showed significant reductions in urine volume, urinary total protein and β2-microglobulin, kidney/body weight ratio, and kidney fibrosis associated with subcutaneous injection of H2."[78]

Diabetes is an inflammatory disease caused by a combination of factors including stress, chemical, and heavy metal toxicity, radiation exposure, magnesium, iodine, and bicarbonate deficiencies, and nutritional imbalances focused on excessive carbohydrate intake that all come together to burn the cellular house down in slow motion. Sugar excess and dehydration create inflammation in the body. This starts a prolonged process of people facing major diseases, including cancer.

"Monitoring of blood-sugar levels, insulin production, acid-base balance, and pancreatic bicarbonate and enzyme production before and after test exposures to potentially allergic substances reveals that the pancreas is the first organ to develop inhibited function from various stresses, writes Dr. William Philpott and Dr. Dwight and Dr. K. Kalita in their book Brain Allergies.

Dr. Lisa Landymore-Lim, in her book Poisonous Prescriptions, explains how many drugs used by the unsuspecting public today are involved in the onset of impaired glucose control and diabetes. It is easy to provoke diabetes in experimental animals with heavy metals like arsenic, mercury, and even fluoride. People are being poisoned with foods full of pesticides, herbicides, preservatives, food additives, and mercury, just to mention a few poisons that plague everyone on earth.

Magnesium is Essential for Diabetics.

Reversing insulin resistance is the first step to reversing diabetes and heart disease. Low serum and intracellular magnesium concentrations are associated with insulin resistance, impaired glucose tolerance, and decreased insulin secretion. Magnesium improves insulin sensitivity, thus lowering insulin resistance. Magnesium and insulin need each other. Without magnesium, our pancreas will not secrete enough insulin–or the insulin it secretes will not be efficient enough–to control our blood sugar.

A Brazilian study published in the journal Clinical Nutrition found that low magnesium levels worsen type 2 diabetes symptoms, resulting in low insulin levels and elevated blood sugar. A diabetic's ability to control blood sugar levels is directly tied to their magnesium levels. The mineral plays a vital role in insulin receptor cells.

Insulin resistance and magnesium depletion resulted in a vicious cycle of worsening insulin resistance and decreased intracellular magnesium. Magnesium is an essential co-factor for enzymes involved in carbohydrate metabolism, so anything threatening magnesium levels threatens overall metabolism.

Over 68% of the U.S. population is magnesium deficient. Up to 80% of *people with type 2 diabetes are low* since they waste more magnesium than others due to out-of-control blood sugar levels. These estimates understate the problem since the tests used to measure magnesium are blood, not cellular levels, of this vital mineral. People with diabetes waste more magnesium due to increased urination from elevated and fluctuating blood sugars, so they constantly need to replenish their magnesium stores.

A Tufts study led by Adela Hruby_found that healthy people with the highest magnesium intake were 37% less likely to develop high blood sugar or excess circulating insulin, common precursors to diabetes. Among people who already had those conditions, those who consumed the most magnesium were 32% less likely to develop diabetes than those consuming the least.

Diabetes and Sodium Bicarbonate.

Parhatsathid Napatalung from Thailand writes, "The pancreas is harmed if the body is metabolically acid. As it tries to maintain bicarbonates, insulin becomes a problem, and hence diabetes becomes an issue. Without the bicarbonate buffer, the disease is far-reaching as the body becomes acid."

The pancreas is a long, narrow gland that stretches from the spleen to the duodenum's middle. It has three main functions. First, it provides digestive juices, which contain pancreatic enzymes in an alkaline solution to provide the right conditions for the digestive process to be completed in the small intestines. Second, the pancreas produces insulin, which controls blood sugar by metabolism and other carbohydrates. Third, it produces bicarbonate to neutralize acids from the stomach providing the right environment for the pancreatic enzymes to be effective.

Allergies start with the body's inability to produce a particular enzyme or to produce enough enzymes for the digestive process to work effectively. In conjunction with this is an inability to produce enough bicarbonate essential for the pancreatic enzymes to function correctly. When this happens, undigested proteins penetrate the bloodstream inducing more allergic reactions. Inflammation is such a scenario; it is systemic but can focus on the pancreas, forcing decreases in bicarbonate production, insulin, and necessary enzymes.[79]

The bicarbonate ion acts as a buffer to maintain the correct acidity (pH) in blood and other fluids in the body. Bicarbonate levels are measured to monitor the acidity of the blood and body fluids. Acidity is affected by foods or medications that we ingest and the function of the kidneys and lungs. The chemical notation for bicarbonate on most lab reports is HCO3- or represented as the concentration of carbon dioxide(CO_2).

The normal serum range for bicarbonate is 22-30 mmol/L. A bicarbonate test is usually performed along with tests for other blood electrolytes. Disruptions in normal bicarbonate levels may be due to diseases that interfere with respiratory function, kidney diseases, metabolic conditions, and a failing pancreas. The pancreas, an organ primarily responsible for pH control,[80] is one of the first organs affected when general pH shifts to the acidic.

Serious Medicine
for the Skin.

After a boiling tea burned a woman's stomach with a first-degree burn. She had been given a hot cup of tea as she was sitting up in bed. She had put lotion on her hands and did not realize they were slippery. It was excruciating, and she was pretty worried because of the large area and pain involved. She said it was not bad at first but got really painful as time went on. She applied hydrogen water to the affected area:

30 minutes later. The redness was gone entirely, and the pain was receding, and she said it then felt "furry" when her shirt would rub against it. That was at 11:00 PM that night. She was able to sleep with no discomfort. The following morning, she had no pain whatsoever.

Sunburn and Hydrogen:

Another time I was all day in the sun and got sunburn on my face, neck, and area around the neck near the shoulders. I have a 4 oz. spray bottle that I infused with H2 for 5 minutes and then sprayed the area several times in 30 minutes. The results were amazing, as the pain dissipated rapidly. After days from the event, here is the result of the experiment:

The H2 infused water:

- Took the pain and discomfort away rapidly.
- DID lessen the pain when you rubbed the skin on the surface.
- DID NOT lessen the pain if you squeezed the skin together (deeper damage?).
- DID NOT lessen the redness/color.
- DID soften the skin.
- My skin DID NOT peel - other than a fine layer - NOT the normal skin peeling off in large patches.
- The skin color is now a normal tan color.
- Wrinkles are going away!

Hydrogen for Beautiful Young Skin.

Famous dermatologist Dr. Nicholas Perricone believes harnessing hydrogen is a far more efficacious way to wage war on wrinkles. A recent Japanese study in the Journal of Photochemistry and Photobiology confirmed what locals had been preaching for years. They found that bathing in hydrogen water every day for three months can reduce neck wrinkles. Similarly, a control group of UV-damaged human fibroblasts (the cell responsible for producing collagen in your skin) was shown to double collagen production after being soaked in hydrogen water for 72 hours.

Molecular hydrogen has been proven to have antioxidant properties, helping remove free radicals from your body that cause premature skin aging. Women and beauty experts are jumping on the hydrogen bandwagon because they can SEE the beautifying effects of hydrogen administration.

We tend to forget that the skin is the largest and most important of human organs. If we can see hydrogen in action on the skin, we can imagine, without scientific studies, that it has a similar effect on internal organs and how they age.

Water is a Hydrogen Rich Substance.

All cells in our body need water to function and stay healthy. If you do not drink adequate water, your cells will suffer from dehydration. Hence drinking hydrogen-rich water, which hydrates better than regular water, will give you moister younger skin.

When thinking of hydrogen gas, either inhaled or dissolved in water, it helps think of water itself, which is obviously a hydrogen-rich substance. When you do not drink enough water, your skin will show dryness, tightness, and flakiness when you do not provide enough hydrogen. The less moisture on your skin, the more it will be prone to wrinkling and age.

Preventing and Treating Strokes with Hydrogen and Magnesium.

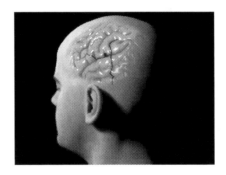

Dr. Matthias Rath believes heart attacks and strokes are not actual diseases but the result of nutritional deficiencies. "Heart disease is not caused so much by what you eat as by what you don't eat."[81] Whatever we choose to believe or conceive, traumatic brain injury (TBI) and cerebral vascular events are usually devastating, yet because few in the mainstream listen to Dr. Rath, heart disease and strokes continue to mow down the public in excessive numbers. In the age of COVID, few even pay attention because of all the hysteria surrounding viruses and vaccines for them.

Annually, in the United States, 795,000 people suffer a stroke, of which 610,000 are first attacks. An estimated 7.2 million Americans aged 20 or more report a history of stroke. Ischemic stroke is one of the most common sources of mortality in the world.

Researchers have tried to find complementary therapy to treat ischemic stroke to improve its prognosis and expand the therapeutic window for reperfusion treatment. There are few treatments proven to lessen brain damage and overall outcome in patients in the mainstream of medicine. But that would change if people listened to Dr. Rath and researched hydrogen gas and high dosages of magnesium.

This chapter is about two neuroprotective agents, which, when brought together, could turn the corner in the prevention of strokes and their treatment. An ideal neuro-protectant would be non-toxic, easily administered, permeable at the blood-brain barrier (BBB), and offer protection at all injury stages, including prevention. Hydrogen gas fits the bill.

Hydrogen Gas is Essential for Stroke Patients.

Every doctor and patient should know that hydrogen has many of the required characteristics for a successful neuroprotectant: it is widely available, is easy to produce, diffuses rapidly through the lipid membranes, is inert and safe to administer, and reacts only with the most aggressive ROS. At the same time, it mediates multiple pathophysiologic pathways leading to apoptosis and cell death.

Hydrogen gas eliminates hydroxyl free radical and peroxynitrite anions and produces a therapeutic effect in patients with ischemic stroke. Many studies have been published illustrating its anti-oxidative, anti-inflammatory, and anti-apoptotic effects.

Numerous experimental evidence indicates that in all forms of stroke damage, the formation of free radicals was increased, leading to nutritional oxidative stress. There are several free radical production mechanisms during ischemia, including intracellular calcium overload, mitochondrial dysfunction, NMDAR-mediated excitotoxicity, and the release of inducible nitric oxide synthase. Excessive free radicals, such as ROS and hydroxyl radical, can damage cellular macromolecules and lead to autophagy, apoptosis, and necrosis of cells by affecting signaling pathways. In addition, free radicals also cause DNA damage and cellular aging.

The key to the success of hydrogen in stroke patients is reducing free radicals as the most critical pathological mechanism of brain damage after stroke. Take down the level of damaging free radicals, and we minimize brain damage. It cannot be highlighted enough that hydrogen gas therapy specifically targets hydroxyl radicals, which are considered the main trigger for free radical chain reactions. One study even showed that hydrogen eye drops directly decreased hydroxyl radicals in ischemia/reperfusion of retinas.[82] Hydrogen can also reduce 8-hydroxy-deoxyguanine, decreasing DNA oxidation.

In a clinical trial of 25 patients with cerebral ischemia, 3% hydrogen was administered by inhalation for one hour twice daily for seven days. Hydrogen concentration reached a plateau at 20 min; then it decreased to 10% after 6–18 min of cessation of administration in arterial and venous blood, respectively. Conclusion: Increasing evidence from cellular, animal and human studies suggests that hydrogen can be administered safely as a neuroprotector during revascularization.

It is time for hydrogen to take center stage for stroke patients. Injections, IV drips of magnesium, and water infused with magnesium bicarbonates would also be called for.

Magnesium for Prevention and Treatment of Stroke.

*Magnesium deficiency can cause metabolic changes
that may contribute to heart attacks and strokes.*

National Institute of Health

When magnesium is present in water, life and health are enhanced. One of the main benefits of drinking plenty of magnesium-rich water is to prevent heart disease and stroke, even in children. Dr. Jerry L. Nadler informs us that magnesium prevents blood vessels from constricting, thus warding off blood pressure rises, strokes, and heart attacks. Magnesium inhibits the release of thromboxane, a substance that makes blood platelets stickier."

A ten-year study of 2,182 men in Wales found that those eating magnesium-low diets had a 50 percent higher risk of sudden death from heart attacks than those eating one-third more magnesium. High magnesium consumers were only half as likely to have any cardiovascular incident such as non-fatal heart attacks, strokes, angina (chest pain), or heart surgery.[83]

A 2006 issue of the *Journal of the American College of Nutrition* published an article showing that as magnesium consumption falls, the level of C-reactive protein goes up. The liver produces C-reactive protein (CRP). It has emerged as a strong predictor of clinical events of cardiovascular diseases, such as heart attacks and stroke, even in cases where cholesterol levels may be expected.

Magnesium & Stroke.

In my practice, the use of magnesium in the early stages of a stroke have rendered the best results.

Dr. Al Pinto

The most effective stroke treatments are given within the first few hours after a stroke has occurred. When magnesium treatments are started within 24 hours from the onset, we see a trend toward a better functional outcome in patients at 30 days versus controls.

Over a decade ago, In Los Angeles, California, we had the FAST-MAG trials, which had the ambulance personal injecting magnesium quickly upon stroke victims' arrival. The Field Administration of Stroke Therapy (FAST-MAG trial) was an NIH-NINDS-sponsored study whose goal is to evaluate field-initiated magnesium's effectiveness and safety in improving the long-term functional outcome of patients with acute stroke.

The FAST-MAG trial addresses the crucial factor of delayed time to treatment, which has hindered all past human clinical trials of neuroprotective drugs.[84] The FAST-MAG Pilot Trial demonstrated that field initiation of magnesium in acute stroke is feasible, safe, and potentially productive. The basic design was to inject magnesium within 1-2 hours of the onset of stroke when the benefits of neuroprotective acute stroke therapies are significant.

By utilizing field delivery via the ambulance, medical scientists conducted the first neuroprotective study ever performed in the 0-2-hour window. Most stroke patients typically do not receive treatment within these brief windows. Patients usually arrive at the hospital too late, with terrible consequences. If ambulances had hydrogen gas machines, we would maximize results.

Researchers believe magnesium slows the chemical process that can kill 12 million brain cells per minute during an untreated stroke, leading to long-term disability and death. Every moment is crucial to the outcome. At least nine pre-clinical studies have examined systemic magnesium sulfate's effect upon final infarct size in animal focal ischemic stroke models. Eight of the nine demonstrated substantial decreases in infarct size in treated animals, with reductions ranging from 26 to 61 percent.

Early studies using rats and mice showed that if given at high concentrations, magnesium can decrease the area of the brain that is permanently lost as a result of a stroke.

Dr. Jose Vega

"How does magnesium protect the injured brain?" asks Dr. Vega. "The response to a lack of oxygen and nutrients (i.e., ischemia) by the brain includes a local release of chemicals that can damage brain cells even beyond the damage that can be expected by ischemia alone. The most harmful of these chemicals is glutamate, an amino acid used in meager amounts by brain cells to communicate. However, during a stroke, the massive amount of glutamate released produces a flood of calcium inside brain cells, which causes them to die prematurely. Magnesium can prevent glutamate from causing this flood of calcium in the cells, thus protecting them from premature death."

Dr. Tavia Mathers and Dr. Renea Beckstrand from Brigham Young University published in the *Journal of the American Academy of Nurse Practitioners* in 2009 that magnesium has been heralded as an ingredient to watch for in 2010 magnesium is helpful for reduction of the risk of stroke.[85]

Dr. Saver and colleagues[86] investigated the neuroprotective effect of early magnesium infusion in ischemic or hemorrhagic stroke in the field; three-quarters of the infarct cohort were treated within two hours of onset and nearly one-third within one hour of onset. Dramatic early results were reported in the early-stage (42 percent of < 2-hour infarct patients) and superior results in the 90-day global functional outcomes (69 percent of all patients and 75 percent of < 2-hour infarct patients).

> *Low CSF Mg+2 levels in patients with*
> *acute ischemic stroke at admission*
> *predicted higher one-week mortality.*[87]

An essential prerequisite for any pharmacological agent to offer significant brain neuronal protection during strokes is its ability to cross the blood-brain barrier freely. Several studies show that magnesium crosses this barrier in both animals and humans.[88] Magnesium ions cross the blood-brain barrier so that intravenous magnesium sulfate significantly raises cerebrospinal fluid and brain extracellular fluid.

Dr. Jerry Nadler said, "Higher dietary intake of magnesium was among the factors associated with a reduced risk of stroke in men with hypertension. In a survey of 45,000 men ages 40-75, the overall risk of stroke was significantly lower for men in the highest quintile of intake of potassium, magnesium, and cereal fiber, but not of calcium, compared with men in the lowest quintile of intake."

Magnesium is an agent with actions on the N-methyl-D-aspartate (NMDA) receptor and has a low incidence of side effects. It may reduce ischemic injury by increasing regional blood flow, antagonizing voltage-sensitive calcium channels, and blocking the NMDA receptor. Systemically administered magnesium at doses that double physiological serum concentration significantly reduces infarct volume in animal models of stroke. A window of up to six hours after onset and favorable dose-response characteristics compared with previously tested neuroprotective agents.[89]

Diabetes, Magnesium, Bicarbonates & Hydrogen.

Most of my writings in my diabetes book centered around the *essential power* of magnesium and bicarbonates working together to manage and even cure diabetes. Early in 2020, at age 67, after a lifetime of overeating sugar, I was diagnosed with diabetes with a fasting level of 132. Fortunately for me, I knew what to do and blew off the diagnosis in two weeks by increasing my magnesium intake.

Magnesium is vital for everyone because it is involved with the creation of insulin, the shape and effectiveness of insulin, and insulin receptivity in the cells. I just ate lunch with lots of honey and jam on my waffles, and my blood tested at only 151, a whole 20 units below normal.

BLOOD SUGAR LEVEL CHART			
	FASTING	JUST ATE	3 HOURS AFTER EATING
NORMAL	80-100	170-200	120-140
PRE-DIABETIC	101-125	190-230	140-160
DIABETIC	126+	220+300	200+

Special Note: Before addressing this chapter's main topic, I want to say that we are at an extraordinary moment. The world is shaking and slipping from under our feet, and that is stressful, exceptionally so. People are dying, suiciding, getting depressed, losing their livelihood, lied to by the media, facing dramatic climate change (to the cold side), facing the collapse of many systems, and a plague of viruses and now vaccines. In terms of medicine, which applies especially so for diabetics, the most important antidote to the stress that ails us is magnesium, and after that, bicarbonates. No matter what else you do, increase your magnesium intake, do not forget about the healing, nurturing sun, and vitamin D intake.

Insulin is a common denominator, a central figure in life, as is magnesium. The task of insulin is to store excess nutritional resources. This system is an evolutionary development used to save energy and other nutritional necessities in times of abundance to survive in times of hunger. Little do we appreciate that insulin is responsible for regulating sugar entry into the cells.

Low serum and intracellular magnesium concentrations are associated with insulin resistance, impaired glucose tolerance, and decreased insulin secretion.[90] Magnesium improves insulin sensitivity, thus lowering insulin resistance.[91] Magnesium and insulin need each other. Without magnesium, our pancreas will not secrete enough insulin–or the insulin it secretes won't be efficient enough–to control our blood sugar.,[92]

White rice, white sugar, white bread, and white pasta are white because they are stripped of their mineral, vitamin, and fiber content. These are poisonous foods for us because they cause magnesium deficiency. It is this magnesium deficiency that is driving up the incidence of diabetes to the pandemic level.

Dr. Carolyn Dean indicates that magnesium deficiency is an independent predictor of diabetes and that people with diabetes need more and lose more magnesium than most people. Magnesium is necessary for the production, function, and transport of insulin.

A study published in the journal *Clinical Nutrition* from a Brazilian researcher's team has found that low magnesium levels worsen type 2 diabetes symptoms, resulting in low levels of insulin and elevated blood sugar. A diabetic's ability to control blood sugar levels is directly tied to their magnesium levels, as the mineral plays a significant role in insulin receptor cells.

Type two diabetes is curable if you ignore your doctor's advice. Diabetes is not the hopeless disease that most doctors would have us believe it is a long-losing battle if you walk the trail western medicine wants you to travel.

Bicarbonate and Diabetes.

Bicarbonate physiology is entirely ignored in diabetes as it is in oncology. Parhatsathid Napatalung from Thailand writes, "The pancreas is harmed if the body is metabolically acid as it tries to maintain bicarbonates. Without sufficient bicarbonates, the pancreas is slowly destroyed, insulin becomes a problem, and diabetes becomes an issue. Without sufficient bicarbonate buffer, the effect of the disease is far-reaching as the body becomes acid."

Understanding sodium bicarbonate begins with a trip to the pancreas, the organ responsible for producing bicarbonate in our bodies. The pancreas is a long, narrow gland that stretches from the spleen to the duodenum's middle. It has three main functions. Firstly, it provides digestive juices, which contain pancreatic enzymes in an alkaline solution to provide the right conditions for the digestive process to be completed in the small intestines. Secondly, the pancreas produces insulin, which controls blood sugar by metabolism and other carbohydrates. Thirdly, it produces bicarbonate to neutralize acids from the stomach to provide the right environment for the pancreatic enzymes to be effective.

Allergies start with the body's inability to produce a particular enzyme or to produce enough enzymes for the digestive process to work effectively. In conjunction with this is an inability to produce enough bicarbonate essential for the pancreatic enzymes to function correctly. When this happens, undigested proteins penetrate the bloodstream inducing more allergic reactions. Inflammation in such a scenario is systemic but can focus on the pancreas forcing decreases in bicarbonate production, insulin, and necessary enzymes.

The bicarbonate ion acts as a buffer to maintain the normal acidity levels (pH) in blood and other body fluids. Bicarbonate levels are measured to monitor the acidity of the blood and body fluids.

There are many causes of diabetes. Heavy metals, toxic chemicals, and radiation contamination will affect, weaken, and destroy pancreatic tissues. When the body is bicarbonate sufficient, it is more capable of resisting the toxicity of chemical insults. That is why the army suggests its use to protect the kidneys from radiological contamination. Much the same can be said for magnesium levels. Magnesium, bicarbonate, and iodine all protect us from the constant assault of harmful chemicals and radiation exposure every day in our water, food, and air.

Hydrogen is Good Medicine for Diabetics.

Hydrogen Medicine is new and was not even on anyone's radar when I wrote *New Paradigms in Diabetic Care*. However, it is now apparent why this most basic gas would help push back against diabetic winds. Molecular hydrogen improves type 2 diabetes by inhibiting oxidative stress. Oxidative stress is recognized as being associated with various disorders, including diabetes, hypertension, and atherosclerosis. It is well established that hydrogen has a reducing action on oxidative stress.

Supplementation of hydrogen water and gas improves lipid and glucose metabolism in patients with type 2 diabetes or impaired glucose tolerance. Hydrogen has been shown to improve glycemic control in Type1 Diabetic animal models by promoting glucose uptake into skeletal muscles. Hydrogen inhalation therapy is also excellent for diabetic neuropathy, no matter where it develops in the body.

Cinnamon.

As a treatment for diabetes, cinnamon gained plausibility in 2003, when a study from Alam Khan suggested several grams of cassia cinnamon per day could lower fasting blood glucose. Khan randomized Type 2 diabetes to 1g, 3g, or 6g of cinnamon for 40 days. All three groups improved their fasting blood glucose and blood lipid levels.

https://www.youtube.com/watch?v=da1vvigy5tQ

In this Ted talk by an obesity specialist, the doctor says that insulin resistance is essentially a carbohydrate intolerance state. What she does not say is that magnesium deficiencies cause carbohydrate intolerance.

What diet is best for diabetes?

Researchers had found a link between higher consumption of white rice and type 2 diabetes. "What we've found is white rice is likely to increase the risk of type 2 diabetes," said Dr. Qi Sun of the Harvard School of Public Health. Those who ate more white rice were 55% more likely to develop the disease than those who ate the least. White rice, white sugar, white bread, and white pasta are white because they are stripped of their minerals (magnesium), vitamin, and fiber content. These are poisonous foods because they cause magnesium deficiency.

A study published in the journal Diabetologia said that a diet of just 600 calories a day reverses diabetes. After one week on a diet, diabetes patients saw their blood glucose levels return to normal, indicating their diabetes had gone into remission. Eight of the eleven patients remained diabetes-free three months after they stopped the diet.

Dr. Lisa Landymore-Lim, in her book *Poisonous Prescriptions,* explains how many drugs used by the unsuspecting public today are involved in the onset of impaired glucose control and diabetes. She explains using the example of the drugs streptozocin and alloxan, which are used in research to make lab rats diabetic.

Powerful Treatments for Parkinson's Disease.

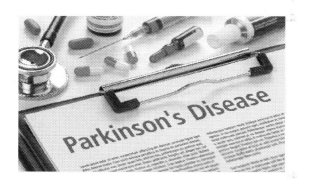

Hydrogen and Magnesium.

"When I first contacted Dr. Sircus to consult with him, I was in desperate need of help. My elderly husband, diagnosed with Parkinson's, suddenly began to decline rapidly over five months. He had no appetite, lost weight, and gradually lost the ability to walk, even with a walker. His legs were swollen from the ankles to the hips, and his mental confusion was getting worse."

"Dr. Sircus prescribed an intensive protocol, which included a hydrogen machine for inhalation of hydrogen gas (AquaCure). He started using the device 5-10 minutes at a time, three times a day, and slowly increasing the time until Dr. Sircus gave the green light to have him hooked up to the gas all day."

"I didn't know what to expect and was careful not to get up my hopes too high, but after about three weeks of inhalation, I could detect a shift for the better. It started with the appetite, which came back vigorously (he ate his breakfast and half of mine too), his legs are no longer swollen, and his old self started to re-emerge. Is he a healthy man yet? Of course not, but I now believe that he will improve his overall health and quality of life with continuous use of hydrogen gas."

"The hydrogen gas has also dramatically improved my chronic ankle pain from an old injury after only two days. The pain reduced by more than fifty percent."

Francisca

The benefits of hydrogen water on motor deficits have been reported in animal P.D. models and P.D. patients.

[93]*Hirayama M, Ito M, Minato T, et al*

One of the most efficient ways to attenuate oxidative stress is drinking hydrogen water and inhaling hydrogen gas. The neuroprotective mechanism of hydrogen reduces lifestyle-related oxidative damage for neurodegenerative diseases, including P.D.

Numerous forms of irreversible damage in nervous system diseases are caused by neuroinflammation, excessive oxidative stress, mitochondrial dysfunction, and cell death.

With gaseous hydrogen H2, one can achieve all kinds of medical and therapeutic results—legions of experts, physicians, biochemists, physicists, and pharmacists have gotten excited about this gas that is sorely lacking in our atmosphere. In a 2015 review by Ichihara et al., there is the mention of "drastic effects," referring to the effectiveness of H2 as a medicine. Such enthusiasm is rare in the otherwise relatively sober research environment.

With the search words "Ohsawa Hydrogen" or "Ohsawa Hydrogen," anyone can quickly get an idea on the Internet of how astonishingly far the topic of hydrogen therapy has already progressed, mainly at renowned universities in Japan, Korea, China, and more recently in the USA.

Parkinson's and Magnesium.

My first experience with Parkinson's fifteen years ago was with a patient of my wife's. I saw her enter the office with much difficulty using a stroller. Her hands shook uncontrollably, and she was miserable. We gave her a little magnesium to take both orally and transdermally. Two weeks later, she came back, but this time she almost ran in without her stroller with the biggest smile I ever saw. Just a little magnesium had done something to her that none of her medications could.

Some people are so magnesium deficient that just supplementing with a little makes all the difference. We are getting to see, during this pandemic, how much pharmaceutical interests control the world of medicine and how much that can hurt people. We can see with our own eyes what concentrated nutritional medicine can do that pharmaceuticals cannot.

Years ago, I went to visit a woman who also has Parkinson's disease. I left her with her iodine, sodium bicarbonate, and magnesium, and she looked like a new woman when I visited again. Her hands barely shook. Today, with the advent of hydrogen inhalation therapy, we can promise Parkinson's patients even more relief. All patients with chronic diseases should employ a complete protocol of natural medicines for the best results. That includes breathing retraining since our breath controls not only physical parameters but also emotional and mental ones.

The National Parkinson Foundation reports that in the United States, 50,000-60,000 new cases of P.D. are diagnosed each year, adding to the one million people who currently have P.D. About four to six million people around the world suffer from the condition.

There is no known cure for Parkinson's disease in contemporary medicine. One main reason is that they refuse to acknowledge how toxicities from heavy metals and chemicals run head-on into nutritional deficiencies, causing illness. Parkinson's and other neurological disorders do respond well to Natural Allopathic Medicine. There are sufficient evidence and testimony to suggest that people do not have to suffer from the worsening disease that leads to total disability. Untreated, Parkinson's leads to a deterioration of all brain functions and an early miserable death![94]

Magnesium, Heavy Metals, and Dopamine.

There is little doubt that low magnesium levels contribute to the brain's heavy metal deposition that precedes Parkinson's, multiple sclerosis, and Alzheimer's. Magnesium protects the cells from aluminum, mercury, lead, cadmium, beryllium, and nickel. Magnesium comes out on top in the class of cerebral protective agents.

Nerve cells use a brain chemical called dopamine to help control muscle movement. Parkinson's disease occurs when the nerve cells in the brain that make dopamine are slowly destroyed. Without dopamine, the nerve cells in that part of the brain cannot properly send messages. This leads to the loss of muscle function. The damage gets worse with time.

It has been shown that continuous low magnesium intake induces complete loss of dopaminergic neurons in rats.[95] Magnesium exerts both preventive and ameliorating effects in an in vitro rat Parkinson's disease model involving 1-methyl-4-phenylpyridinium (MPP+) toxicity in dopaminergic neurons.[96]

Magnesium protects dopaminergic neurons in the substantia nigra from degeneration. There is a significant and striking effect of magnesium for preventing neurite and neuron pathology and amelioration of neurite pathology.[97]

Magnesium Testimonial.

I had a beautiful conversation with Karin, who is taking care of her father, Peter, who has had Parkinson's for over 18 years. His condition before starting the magnesium oil was that he could not talk at all. Could not articulate what-so-ever! He was barely functional and did nothing voluntarily. The drooling was getting so bad and constant that Karin was beginning to isolate him in his bedroom; it worsened by the week. He also had started getting violent with her.

She applied the magnesium oil twice, and he woke the following morning, washed his face, cleaned his teeth, and put on his robe by himself—without being told. This did not happen for two years. What is more, he was not drooling.

After only three days, his speech has been better overall. Karin applied magnesium faithfully three times a day all over him. His eyes are brighter, the concentration is more prolonged and better, and the speech is much improved. At least he can string two or three words together now and does not freeze up completely.

Since starting the magnesium oil, his demeanor has improved immensely. There are no more surly ugly looks, no more stubborn refusals to swallow or do something that she asks him to do. Excellent improvement, and best of all, he can now communicate to tell her what he wants and needs.

Cannabis & Parkinson's Disease.

Cannabinoids are potent antioxidants that can protect neurons from death without cannabinoid receptor activation. It seems that cannabinoids can delay or even stop the progressive degeneration of brain dopaminergic systems.

Dr. Evzin Ruzicka, an attending neurologist at Charles University in Prague in the Czech Republic, said, "To our knowledge, this is the first study to assess the effect of cannabis on Parkinson's disease, and our findings suggest it may alleviate some symptoms.[98]

In a 2007 study published in *Nature*, researchers from the Stanford University School of Medicine report that endocannabinoids, naturally occurring chemicals found in the brain like the active compounds in marijuana and hashish, helped trigger a dramatic improvement in mice with a condition similar to Parkinson's.

Dr. Robert Malenka, the Nancy Friend Pritzker Professor in Psychiatry and Behavioral Sciences, and Dr. Anatol Kreitzer combined a drug used to treat Parkinson's disease (dopamine) with an experimental compound that can boost the level of endocannabinoids in the brain. When they used the combination in mice with a condition like Parkinson's, the mice went from frozen in place to freely in 15 minutes. "They were basically normal," Kreitzer said.

Iodine.

Iodine is found in large quantities in the brain and the ciliary body of the eye. Lack of iodine may be involved in production of Parkinson's disease and glaucoma.

Dr. James Howenstein

Long-term iodine deficiency appears to be linked to abnormalities in the dopaminergic system, including increased dopamine receptors.

The hypothesis that Parkinson's disease may be linked to soil and dietary iodine deficiency associated with glaciation is not new.

> *In the brain, iodine concentrates in the*
> *substantia nigra, an area of the brain that*
> *has been associated with Parkinson's disease.*

> *Dr. David Brownstein*

Hydrogen Dosages.

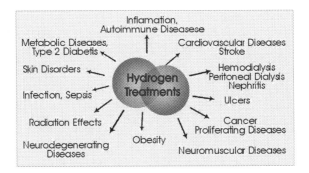

Several doctors are experiencing good results - both personally and with their patients offering 30-60 minute sessions 2-3 times a week with the inhalation machine and then working with the VR hydrogen tablets for their daily dose when not breathing H2. The key to understanding hydrogen supplementation dosages is to be taken over the long term, not for a "just a month and you're cured" type of mentality.

I suggest one tablet on an empty stomach morning, noon, and night for the best results. Hydrogen dosages should be compared to taking daily vitamins, meaning they should be incorporated into a daily routine that becomes permanent.

Higher dosages are needed for anyone seriously ill or with any chronic disease. When one needs to dig out of a dark pit of medical despair, hydrogen will provide the muscle along with high dosages of magnesium (80 mg included with every tablet) and just the right amount of bicarbonate to raise pH to a more alkaline state. Of course, this all changes if one uses a hydrogen inhaler, enabling patients to take in hydrogen at higher dosages.

The newest generation of hydrogen tablets offers hydrogen dissolved in water at very high parts per million. I use slightly icy water and add one tablet to a very large glass of water. As soon as the tablet dissolves, I down the water to get the most hydrogen. If your water is very cold, it takes a little longer. To use more than one tablet is a waste of money.

If one puts the tablet in a tightly closed bottle, then the water temperature does not matter. I like higher doses of H2 taken at once. Higher dosages are in order after heavy exertion, air travel, lack of sleep, intense exercise, stress, etc. Usually, I would recommend taking one tablet two or three times a day when in good health.

If someone is dying or suffering from late-stage cancer, hydrogen is vital. Just hook the person up to a machine and keep breathing until improvement is noticeable. Hydrogen can also be inhaled while sleeping.

General Information on Dosages.

The dosage determines the effect when using nutritional medicines like magnesium chloride, iodine, sodium bicarbonate, vitamin C, alpha-lipoic acid, iodine, selenium, and hydrogen. In conventional allopathic medicine, they say the dose makes the poison, but we do not use poisons in the Natural Allopathic Medicine protocol. Now I often call this medical model Innovative Medicine.

In Innovative Medicine, we usually take a dosage to exceedingly high levels without the side effects found in pharmaceuticals that are an ever-present danger even at very low doses. This general information on dosages is very applicable to hydrogen.

Even water and vitamin C are placed on a toxicity scale in allopathic medicine, with everything being defined as poisonous. Though one can indeed drown in water, a large person can safely drink a gallon of it without ill effect. And one can put pounds of magnesium chloride in a bath and take high doses of iodine safely for infectious disorders without the severe and dangerous downside of antibiotics. Adverse effects are rare and usually attributed to a lack of care or knowledge of the person or prescriber.

It is undoubtedly possible to cure incurable diseases by using the correct doses of vitamins, minerals, and fatty acids (among other things). The quantity determines the effect! Low doses do not get clinical results! Through the years, I have seen the mistake of under-dosing over and over.

Cardiologist Dr. Thomas Levy said, "The three most important considerations in effective vitamin C therapy are dose, dose, and dose. If you do not take enough, you won't get the desired effects."

Effective doses are high quantities, often hundreds of times more than the U.S. Recommended Dietary Allowance (RDA) or Daily Reference Intake (DRI). Dr. Abram Hoffer said, "Drs. Wilfrid Shute and Evan Shute recommended doses from 400 to 8,000 IU of vitamin E daily. The usual dose range was 800-1600 IU, but they report that they had given 8,000 IU without seeing any toxicity." The Shutes successfully treated over 35,000 patients with vitamin E.[99]

If one is lucky enough to have a hydrogen inhaler and is in critical condition, one does not worry about the dosage. Four hours a day or longer should make a significant impact on severe conditions.

Critical Dosages.

Knowing appropriate dosages is essential to practitioners and patients because dosages are mission-critical for achieving therapeutic effects. We all must estimate dosages. The most basic thing we do is our water intake. What is the right amount of water (dosage) a person should drink in a day? What dosage of the sun? These are important questions that doctors often answer incorrectly.

In pharmaceutical medicine, the principle of measuring a critical dose of medicine would be to measure the minimum necessary therapeutic effect. That is a critical dosage because what comes with increasing dosages of synthetic chemicals (all pharmaceuticals are mitochondrial poisons) is an increase in side effects and an increased chance of dying from prescribed medicines.

In her paper published in JAMA in July 2000, Dr. Barbara Starfield reported that 106,000 people in America alone die from properly prescribed pharmaceutical medicines. In 2015, 443,900 total deaths from prescription medication were reported to one of the US's poison centers, with analgesics—or painkillers—being the deadliest poison. Of these, 275,000 were due to errors—such as a wrong dose—and 130,000 were caused by unintentional misuses, such as taking the drug more frequently than prescribed. Forty thousand deaths were attributed to an adverse reaction to a drug that was properly prescribed and taken.[100]

More than 75,000 ER visits a year due to Tylenol overdoses causing severe health emergencies, including liver failure and death. There has not been a person who has died from an iodine overdose in 50 years, yet plenty of doctors are afraid to prescribe iodine. There are no such things as safe drugs. "Safe" Pharmaceutical Poisons Do Not Exist, but there is a host of safe but powerful natural substances that Mother Nature gives us to heal ourselves.

The top five causes of poisoning in a recent study were,
in order, antidepressant medications, analgesics such
as aspirin, street drugs, cardiovascular drugs, and alcohol.

Royal Society of Chemistry

That is what happens when you deal with dangerous drugs. After 400 years, modern medicine has a problem with its own paradigm, "The dose makes the poison." One of the reasons is the art, science, and absurdity of using poisons as medicine has been clouded because most city dwellers are already filled with toxins from the air and water they are exposed to. Adding more poisons to a person's already sky-high toxic burden does not work.

The dose that makes the poison is a long-accepted concept, but medicine is supposed to be about helping people. The word poison was first recorded in Middle English in a work composed around AD 1200. When introduced into or absorbed by a living organism, a toxin is any substance that destroys life or injures health. Poison is defined as any substance capable of producing a morbid, toxic, or deadly effect. A poison is a material that inhibits other substances, especially enzymes, and the vital biochemical processes in which they are involved. Enzymes are crucial because every chemical change to repair tissue or assimilate food involves enzymes' activity. Without enzyme activity, there is no biological activity, no life.

Poisons' effects can be quick or extremely slow, building gradually up, creating low-grade debilitation diseases like chronic fatigue syndrome or devastating neurological disorders like MS, ALS, and Alzheimer's. Nothing will burn up a neuron faster than heavy metal mercury injected into babies starting at six months of age. Vancouver scientist Dr. Chris Shaw shows a link between the aluminum hydroxide used in vaccines and symptoms associated with Parkinson's, amyotrophic lateral sclerosis (ALS, or Lou Gehrig's disease), and Alzheimer's.[101]

Wrong Dosages Wrong Medical Concepts.

If the dose is low enough, even a highly toxic substance will cease to cause a harmful effect. Thus, the toxic potency of a chemical is defined by the dose (the amount) of the chemical that will produce a specific response in a specific biological system. "In all of these debates, the key point that is not often understood is that it's the dose that makes the poison," says Dr. Carl Winter, an expert in toxicology at the University of California, Davis. "The tendency is to exaggerate toxicity. It's a slippery slope where to draw the line on what represents a legitimate concern and what restrictions should apply."[102] This is, of course, not true for something as radical as plutonium. The crucial issue with plutonium is not volume or mass—it's toxicity.

The Romans were aware that lead could cause serious health problems, even madness, and death. However, they were so fond of its diverse uses that they minimized the hazards it posed. What they did not realize was that their everyday, low-level exposure to the metal rendered them vulnerable to chronic lead poisoning, even while it spared them the full horrors of acute lead poisoning. Roman engineers brought down the Roman Empire when they replaced their stone aqueducts with lead pipes to transport and supply drinking water, turning much of the Roman population into neurological cripples.[103]

"The prolonged effects of low-grade concentrations of toxic substances depend on individual susceptibility," says Professor I. M. Trakhtenberg from the former Soviet Union. The science of low-level toxicity shows that it matters on the parts per million, per billion, and even per trillion level. As our instruments have become infinitely more sensitive, scientists have been able to penetrate new worlds of chemical sensitivity that Paracelsus could not have imagined. What industry and government have hidden in the low numbers seen in parts per million becomes astronomical when calculated and plotted as parts per trillion.

Hormesis.

"What doesn't kill you makes you stronger" is a phrase that many feel contains more than a grain of truth. It describes hormesis theory — the process whereby organisms exposed to low levels of stress or toxins become more resistant to tougher challenges. It is a cousin concept for "The dose makes the poison."

In recent years, biologists have pieced together a clear molecular explanation of how hormesis works; thus, the theory has finally been accepted as a fundamental biology and biomedicine principle. An example of this principle is seen when exposing mice to small gamma-ray radiation doses shortly before irradiating them with very high gamma rays levels. This decreases the likelihood of cancer. A similar effect occurs when dioxin is given to rats.

A low dose of poison can trigger specific repair mechanisms in the body. These mechanisms, having been initiated, are efficient enough to neutralize the toxin's effect and even repair other defects not caused by the toxin. It's a nice theory to justify poisoning people and taking a potluck at achieving precisely the correct dose. It makes much more sense to dance with magnesium and other extraordinarily safe substances that do not poison the patient.

Calculating Dosages for Natural Medicine.

In natural medicines, the critical dosage is measured by the maximum that can be taken to achieve the needed therapeutic effect. In *Natural Allopathic Medicine,* we often take doses to exceedingly elevated levels without the side effects found in pharmaceuticals that are an ever-present danger even at very low doses.

The secret to safe and effective medicine is using medicinal substances that do not have side effects in reasonable doses, meaning they are not poisons. This is the very meaning of safe - something that will not harm or hurt you. The key to any natural protocol is getting the doses high enough. It is best to start low with most substances, get used to each substance, and slowly bring them up.

Conclusion.

Aurelius Phillipus Theostratus Bombastus von Hohenheim, immortalized as "Paracelsus" and sometimes called "the father of toxicology," was born in 1493. Paracelsus, a Swiss doctor, pioneered the use of chemicals and minerals in medicine. Paracelsus was the first to say, "It depends only upon the dose whether a poison is a poison or not. A lot kills; a little cures." So he would take a very toxic substance like mercury and use it to cure epilepsy, something no one in his right mind would do now.

The assumption that poisons can be used safely is modern man's Pandora's Box; once opened, the greediest power-hungry industrialists felt free to use poison in everything from household products, like soap and shampoo, to putting it directly in our foods, medicines, and waer.

Hydrogen Water

- Drinking water that contains hydrogen is an effective way to take hydrogen into the body. The hydrogen present in the water is proven to increase hydration levels, proven to satiate your thirst six times more than tap water.
- Hydrogen water can improve the absorption of supplements taken. It can also help absorb nutrients from food much better than regular tap water.
- It can help lubricate body joints and muscles. Hydrogen-rich water provides around 70% lubrication to joints and muscles.
- Hydrogen water can help the brain perform better. Our mind cannot use its full potential without proper hydration.
- It can help remove impurities inside your body. It can also improve the quality of your cells and organs.
- Drinking hydrogen water can help regulate your blood pressure to maintain optimum performance.

- It reduces the symptoms of some chronic illnesses known to attack the body. It can also help prevent the onset of symptoms of arthritis and relieve headaches.
- It hydrates the skin. Dehydration can make skin flaky and sag in time. Proper hydration is needed to maintain skin quality.
- It reduces damage taken from the sun and wind. It also helps hydrate hair follicles.

Drinking dissolved hydrogen.

The concentration/solubility of hydrogen in water at standard ambient temperature and pressure (SATP) is 0.8 mm or 1.6 ppm (1.6 mg/L). For reference, conventional water (e.g., tap, filtered, bottled, etc.) contains less than 0.0000002 ppm of H_2, which is well below the therapeutic level. The concentration of 1.6 ppm is easily achieved by many methods, such as merely bubbling hydrogen gas into water. Because of the molecular hydrogen's low molar mass (i.e., 2.02 g/mol H_2 vs. 176.12 g/mol vitamin C), there are more hydrogen molecules in a 1.6-mg dose of H_2 than there are vitamin C molecules in a 100-mg dose of pure vitamin C (i.e., 1.6 mg H_2 has 0.8 millimoles of H_2 vs. 100 mg vitamin C has 0.57 millimoles of vitamin C).

The half-life of hydrogen-rich water is shorter than other gaseous drinks (e.g., carbonated or oxygenated water), but therapeutic levels can remain for a sufficiently long enough time for easy consumption. Ingestion of hydrogen-rich water results in a peak rise in plasma and breath concentration in 5-15 min, depending on the dose.

The rise in breath hydrogen is an indication that hydrogen diffuses through the submucosa and enters systemic circulation, where it is expelled out of the lungs. This increase in blood and breath concentration returns to baseline in 45-90 min depending on the ingested dosage.

Water as a Medicine.

Water is oxygen and hydrogen. Water is a wonderful medicine so necessary to life. We know drinking water frequently helps protect individuals from chronic kidney disease (CKD). At the Canadian Society of Nephrology's 2013 annual meeting, researchers reported that CKD was 2.5 times less likely to develop in people who drank more than 4.3 liters of water a day than those who drank less than two liters a day.

"Water, the Hub of Life. Water is its mater and matrix, mother and medium. Water is the most extraordinary substance! All its properties are anomalous, enabling life to use it as a building material for its machinery. Life is water dancing to the tune of solids," wrote Albert Szent-Gyorgyi. It is hydrogen and oxygen that dancing.

Dr. Gerald Pollack, professor of bioengineering, received the highest honor that the University of Washington at Seattle in the United States could confer on its staff for his work with water. Professor Pollack says, "Water covers much of the earth. It permeates the skies. It fills your cells — to a greater extent than you might be aware. Your cells are two-thirds water by volume; however, the water molecule is so small that if you were to count every single molecule in your body, 99% of them would be water molecules. That many water molecules are needed to make up the two-thirds volume. Your feet tote around a huge sack of mostly water molecules."

Hydrogen researcher George Wiseman reminds us of introductory chemistry, saying, "By weight, our bodies are mostly oxygen because oxygen is eight times heavier than hydrogen. By volume, our bodies are mostly hydrogen because there are two hydrogen atoms for every oxygen atom." Hopefully, the message is getting clear. When we employ hydrogen and oxygen, we practice a supreme life-extending health practice that doubles as medicine in the diverse forms available.

We all know that dehydration is a problem that causes suffering and disease. Dehydration is a deficiency of molecules of hydrogen bonded to oxygen. Oxygen deficiency is a problem, and so is an excess of charged hydrogen atoms, which drives the blood and body into acidity.

For about two years, I have been experimenting with different molecular hydrogen products. I have recently been using a new generation of tablets that are the best, most convenient highest ppm. There are several labels and types of hydrogen water tablets.

The most distinguished players in this market have been Vital Reaction; Active H2, one of the original players, has also developed a popular new hydrogen tablet. The first player in the hydrogen tablets was Megahydrate, which works differently than molecular hydrogen.

The Vital Reaction has elevated levels (ppm at 10 or above) and the highest magnesium, which registers 80 mg per tablet. Active H2 Ultra offers around 8 ppm for a slightly lower price, and there is no residue in the glass, making for pleasant hydrogen consumption.

None of the hydrogen water machines on the market can create hydrogen at the high levels that the newest generation of tablets can, which are exceptional in hydrogen concentration. They provide far more hydrogen in the shortest amount of time. Moreover, they make their hydrogen in an open container for convenience and effortlessness of use.

Hydrogen tablets are great for all first-time users. Even if one purchases a dedicated hydrogen gas inhaler that simultaneously makes hydrogenated water, the tablets offer portability and more potency. When one's medical needs are critical, one wants to jump on as high a dose as possible of hydrogen.

"I've been using Vital Reaction tablets daily for the past four months and have been noticing multiple positive changes in my health. While I am generally in excellent health and take no prescription medications, I have been battling brain fog for several years. This negatively impacted my work, academic efforts, and personal life as I could not recall information quickly when needed. Since starting to drink hydrogen water, I have been able to think more clearly, and my memory recall has sharpened. I have also noticed that my energy level has increased since using the tablets. I no longer get tired during the day and have also been sleeping more soundly and not waking during the night. Finally, previous pain in my muscles has seemed to dissipate. I do not have any more body pain. I love my hydrogen water and am committed to continuing it as a part of my daily routine," says Dr. Leslie Nye.

Hydrogen Oxygen Needed By All Seriously Ill COVID Patients.

There are so many factors contributing to COVID deaths we need a full-length book to talk about it. Starting with masks (muzzle, suffocation, and pollution devices), we know that they will cut down on oxygen delivery to the cells when used extensively.

Hydrogen Medicine holds a significant promise for all patients, including those with viral infections. Early on in the pandemic, Chinese doctors applied a viable approach to treating COVID patients already in ICU—inhaling hydrogen and oxygen (Brown's Gas). It is an innovative, environmentally friendly treatment with no side effects involving the electrolysis of H_2O to produce a mixture of hydrogen and oxygen for human inhalation.

The research results as a treatment for COVID-19 were published in the June 2020 edition of the Journal of Thoracic Disease (JTD). In a multicentered, open-label clinical trial described in the article entitled "Hydrogen/oxygen mixed gas inhalation improves disease severity and dyspnea in patients with Coronavirus disease 2019 in a recent multicenter, open-label clinical trial". Inhalation of the hydrogen/oxygen mixture prevented further disease progression and, most notably, alleviated the shortness of breath experienced by COVID-19 patients.

This is a retrospective study, which validated the efficacy and safety of inhaling an H_2-O_2 mixture in patients with COVID-19. The research and clinical trial associated with the study were led by Zhong Nanshan, China's leading pandemic control expert and head of the National Health Commission's expert group fighting COVID-19.

The clinical trial included 44 patients with COVID-19 whose dyspnea could not be alleviated after regular treatment in 11 hospitals. After three days of continuous use of the hydrogen/oxygen, all 44 patients showed significant improvement compared with 46 patients who received the standard regimen of oxygen alone.

The Ultimate in Lung Therapy.

Patients who inhaled the hydrogen-oxygen gas mix showed improvement in chest pain symptoms, dyspnea, shortness of breath, cough, sputum, and violent pneumonia in patients. It can also reduce the risk of severe illness, shorten hospital stay length, and help patients improve pulmonary fibrosis.

Asclepius' director Xin-Yong Lin said, "We believe this is an extremely innovative treatment tool without any side effects, which is a result that all hydrogen medical experts have been looking forward to," he said. "This is undoubtedly inspiring given the current lack of drugs for the novel coronavirus pneumonia."

Hydrogen Medicine
for the Flu.

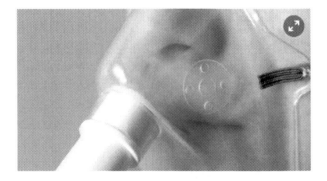

The CDC (Centers for Disease Control) never learns because they do not want to. And in the middle of the flu season, people pay with their lives. Every year hundreds of people needlessly die during flu season.

The CDC has called the 2017-2018 flu an epidemic. In England, it was a pandemic filling up hospitals with standing rooms only. They reported that more than 50 patients at a time had been left waiting for beds in casualty units, with 120 patients a day in corridors and "some dying prematurely."

On the 12th of January that year, the CDC announced a first in its 13 years of flu monitoring: As of Jan. 6, every part of the continental US showed "widespread" flu activity. The flu was so widespread that the agency has declared it an epidemic and urged those who have not been vaccinated to seek out the flu shot. This is a bad idea, not only because it has neuro-toxic mercury but also at best it is only 30 percent effective, according to CDC officials.

"This is the first year we had the entire continental US be the same color on the graph, meaning there's widespread activity in all of the continental US at this point," CDC Influenza Division Director Dr. Dan Jernigan said during a Friday briefing. "It is in a lot of places and causing a lot of flu." At that time, the hospitalization rate was 22.7 people per 100,000 U.S. residents.

While causing mild disease in most people, the Flu can also cause severe illness and death in others. The flu may also exacerbate existing chronic conditions, particularly among older adults, leading to complications and death.

If you see a doctor within 48 hours of developing these symptoms, you may be able to take antiviral medications, such as Tamiflu, which may shorten the course of the illness by only one day. Tamiflu is and has always been a big mistake!

As soon as flu vaccinations start next month, some people getting them will drop dead of heart attacks or strokes, some children will have seizures, and some pregnant women will miscarry. – New York Times 2009.[104] You would think the public media and the CDC would insist that people know that.

Living with cancer increases your risk for complications from influenza ("flu"). If you have cancer now or have had cancer in the past, you are at a higher risk of seasonal flu or influenza complications, including hospitalization and death. This is especially true if you are having surgery, chemotherapy, or radiation. "People tend to underestimate the effects of flu; it puts several hundred thousand people in the hospital every year. It can kill up to 50,000 people every year, and there are certain groups—people with underlying medical conditions, the elderly and the young—who are at risk for severe illness from flu," said Dr.

Michael Jhung, a flu expert at the U.S. Centers for Disease Control and Prevention.

Replacing the Need for Flu Vaccine and Tamiflu.

Hydrogen gas, especially when administered with oxygen, will save the day in medical centers and at home too, when one has severe flu. It is hard to die of the flu if you are hooked up to a hydrogen inhaler. Keeping warm with an infrared mattress is especially comforting for those who have a severe case of the flu.

Death via Cytokine Storms.

We need to understand why hydrogen is the contest winner for the best flu medicine. A cytokine storm (what kills the patient with influenza) is high fever, swelling, redness, extreme fatigue, and nausea. In some cases, the immune reaction may be fatal. To stop the cytokine storms and acute respiratory distress syndrome (ARDS), we have to turn away from the orthodox medical responses using vaccines and drugs like Tamiflu.

Data from clinical trials involving patients with sepsis-induced ARDS have shown a reduction in organ damage. A trend toward improving survival (survival in ARDS is approximately 60%) after administering various free radical s.cavengers (antioxidants).

Hydrogen to the Rescue.

Recent studies revealed that intraperitoneal injection of hydrogen-rich saline has surprising anti-inflammatory, anti-oxidant, and anti-apoptosis effects. It also protected the organism against polymicrobial sepsis injury and acute peritonitis injury by reducing oxidative stress and decreasing mass pro-inflammatory responses. It is well known that most viral-induced tissue damage and discomfort are mainly caused by an inflammatory cytokine storm and oxidative stress rather than by the virus itself.[105]

Studies have shown that suppressing the cytokine storm and reducing oxidative stress can significantly alleviate the symptoms of influenza and other severe viral infectious diseases. Thus, medical scientists hypothesize a hydrogen-rich solution therapy to be a safe, reliable, and effective treatment for Multiple Organ Dysfunction Syndrome (MODS) induced by influenza and other viral infectious diseases.

Combination Therapy for the Flu.

Of course, other excellent therapies avoid the flu's worst symptoms and pains, with infrared therapy being at the top of the list. Jumping on a BioMat when suffering from the flu is like jumping into heaven. The only place to be during a flu attack; crashed out on a BioMat! The best approach is to administer hydrogen and oxygen gases with an inhaler combined with hydrogen water for complete hydration in terms of hydrogen delivery.

Natural Allopathic Emergency Medicine for Flu.

Five years ago, I published a protocol against influenza that involves basic emergency room and intensive care medicines that are an integral part of my medical approach. It includes Magnesium chloride, iodine, sodium bicarbonate, selenium, vitamin C, and vitamin D3 (if possible, through sun exposure) as medicines that help patients resist and even avoid dangerous flu complications.

Breathing retraining (slow breathing) aids in increasing CO_2 and O_2 levels, and of course, complete hydration therapy is essential. Hydrogen is the perfect medicine to mitigate the worst effects of the flu. Molecular hydrogen can significantly down-regulated expressions of inflammation-related genes and selectively reduce hydroxyl radical and Peroxynitrite.

Infrared Therapy – Too Good to be True for the Flu.

If it is possible to love a medical device, one will fall for the infrared BioMat. It keeps you warm on cold nights and takes away the pain. This is especially true if one is suffering from the flu. Everyone deserves a BioMat, just as everyone deserves love. I call them medical love machines. BioMats are that good and that helpful. I wish everyone could afford one!

When far-infrared (FIR) heat penetrates through the skin to the subcutaneous tissues, it transforms from light to heat energy, dilates blood capillaries, and assists the body in eliminating toxins and metabolic wastes sweating. Activated by heat, the FIR energy is absorbed by human cells in a process known as "resonance" or "resonant absorption."

The adaptive capacity of a cell determines its fate when it comes to stress. Far infrared treatments reduce the cells' pressure by nourishing them with light and heat, increasing the nutritional sufficiency of oxygen, and increasing cellular respiration. More toxins and wastes leave the cells.

This device is best for elevating internal body temperature to melt away old waste products, increase circulation, and reduce and eliminate pain. It is only a couple of inches thick, designed to put on a firm mattress or the floor. You can adjust the temperature to your liking, and it is the most wonderful feeling in the world.

It is different from an ordinary electric heating pad. The surface is not warm to the touch, but you feel warm on the inside when you lie down on it. It's a lying-down sauna, and if you turn the settings up, it will make you sweat without you having to move a muscle. You can sleep on it all night!

Far infrared heat is beneficial to people in many ways:

1. FIR heat expands capillaries, stimulating increased blood flow and aiding in regeneration, improved circulation, and oxygenation.
2. Far infrared heat speeds cellular metabolic rates by stimulating mitochondria and triggering enzyme activity, promoting the destruction of pathogens: bacteria, viruses, fungi, and parasites. FIR energy strengthens the immune system by stimulating the increased production of white blood cells (leukocytes) by the bone marrow and killer T-cells by the thymus gland.
3. Far infrared heat promotes the rebuilding of injured tissue by positively affecting the fibroblasts (connective tissue cells necessary for repair). It increases the growth of cells, DNA, and protein synthesis critical during tissue repair and regeneration.

Sodium Bicarbonate (Baking Soda).

"In 1918 and 1919, while fighting the 'Flu' within the U. S., rarely anyone who had been alkalinized with bicarbonate of soda contracted the disease. Those who did acquire it, if alkalinized early, would invariably have mild attacks. I have since that time treated all cases of 'cold,' influenza, and 'la gripe' by first giving generous doses of bicarbonate of soda, and in many, many instances within 36 hours, the symptoms would have entirely abated." (Dr. Volney S. Cheney to the Arm & Hammer Company.)

Vitamin D Will Protect Against the Flu.

One of the most significant influenza triggers, the swine flu, and deaths from pulmonary deficiency is vitamin D deficiency. Vitamin D reduces the risk of dying from all causes, including the flu. Researchers from Winthrop University Hospital in Mineola, New York, found that giving vitamin D supplements to a group of volunteers reduced infection episodes with colds and flu by 70% over three years. The researchers said that the vitamin stimulated "innate immunity" to viruses and bacteria.

Iodine.

"Extremely high doses of iodine can have serious side effects, but only a small fraction of such extreme doses are necessary to kill influenza viruses," writes Dr. David Derry of Canada. In 1945, a breakthrough occurred when J. D. Stone and Sir McFarland Burnet (who later went on to win a Nobel Prize for his Clonal Selection Theory) exposed mice to the lethal effects of influenza viral mists. The deadly disease was prevented by putting iodine solution on mice snouts just prior to placing them in chambers containing influenza viruses. Dr. Derry reminds us that students in classrooms were protected from influenza a long time ago by iodine aerosol therapy. Aerosol iodine also is effective against the freshly sprayed influenza virus.

Selenium.

Selenium is a strong antioxidant anti-inflammatory. Vascular surgery is made safer; reperfusion injury, myocardial infarction, and ischemic stroke are alleviated with selenium injections, as are cytokine storms provoked from out-of-control infections.

Clinical investigations in sepsis studies indicate that higher doses of selenium are well tolerated. Continuous infusions of selenium as sodium selenite (4,000 µg selenium as sodium selenite pentahydrate on the first day, 1,000 µg selenium/day on the nine following days) had no reported toxicity issues. Given this information, Biosyn, a pharmaceutical company, introduced the 1,000 µg dose vials for such high selenium clinical use.

Selenium-deficient mice developed much more severe lung pathology after infection with influenza virus than did selenium-adequate mice. In another study, when selenium-deficient mice were infected with a mild strain of influenza virus, the virus mutated to become a more virulent strain, one that caused severe lung pathology even in selenium-adequate mice.

Selenium is also an antidote to mercury, having a higher affinity for it than any other atom. So selenium does double duty, lowering mercury toxicity. This is important because mercury toxicity is known to provoke influenza.

Magnesium Chloride.

Magnesium chloride (magnesium oil) has always been and remains my favorite medicine that affects overall physiology positively and directly. Dr. Raul Vergini says, "Magnesium chloride has a unique healing power on acute viral and bacterial diseases. It cured polio and diphtheria and is the main subject of my magnesium book. A few grams of magnesium chloride will clear nearly all acute illnesses every few hours, which can be beaten in a few hours. I have seen a lot of flu cases healed in 24-48 hours, with three grams of magnesium chloride taken every 6-8 hours."

Intravenous Vitamin C.

Intravenous vitamin C is an intense treatment when people are between life and death. It has the power to bring people back from the brink. Vitamin C (ascorbic acid) contributes to a wide range of benefits. It performs many critical functions within the body involving detoxification, tissue building, immune enhancement, pain control, and controlling or killing pathogenic organisms. It is also known to help wound and bone healing, healthy skin and eyes, fighting infections, stress control, toxic exposure, and repairing damaged tissue of all types.

Cannabinoid Medicine.

At the University of Reading pharmacy department, Dr. Ben Whalley said tests in animals had shown marijuana compounds to effectively prevent seizures and convulsions while also having fewer side effects than existing epilepsy drugs. At the National Institutes of Health (NIH) in Bethesda, Md., rat nerve cells were exposed to a typically released toxin during strokes. Cannabidiol reduces the extent of damage, reported the National Academy of Sciences. More effective than vitamins C or E, potent antioxidants such as cannabidiol (CBD) will neutralize free radicals without the accompanying high experience with marijuana used for recreational and other medical purposes. All forms of cannabis have anti-oxidative, neuroprotective, immunomodulation, analgesic, and anti-inflammatory actions.

Beyond these core physiological protective mechanisms, something as simple as smoking marijuana is ideal for influenza's pain and discomfort. With or without the "high," cannabinoid medicine offers safe pain relief even as it heals and protects. It should be put into wide use in hospitals and at home for routine treatment against the flu's worst ravages.

What I am saying for adults above also applies to children. Dr. Ester Fride strongly recommends the use of cannabinoids in pediatric medicine. She notes that "excellent clinical results" have been reported in pediatric oncology and case studies of children with severe neurological diseases or brain trauma. She suggests that cannabis-derived medicines could also play a role in treating other childhood syndromes, including the pain and gastrointestinal inflammation associated with cystic fibrosis.

Strengthening the Immune System.

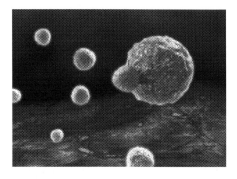

A white blood cell, also known as a T cell, carries unique structures on its surface to recognize specific pathogens.

"On the whole, your immune system does a remarkable job of defending you against disease-causing microorganisms. But sometimes it fails: A germ invades successfully and makes you sick. Is it possible to intervene in this process and make your immune system stronger? What if you improve your diet? Take certain vitamins or herbal preparations? Make other lifestyle changes in the hope of producing a near-perfect immune response? The idea of boosting your immunity is enticing, but the ability to do so has proved elusive for several reasons. The immune system is precisely that — a system, not a single entity. To function well, it requires balance and harmony. There is still much that researchers don't know about the intricacies and interconnectedness of the immune response." Harvard Health.

The type of immune therapy the pharmaceutical companies and oncologists support costs upwards of a million dollars a year. Nivolumab costs $28.78 per mg, whereas Ipilimumab costs $157.46 per mg. "To put that into perspective, that's approximately 4000 times the cost of gold," commented Leonard Saltz, MD, from Memorial Sloan Kettering Cancer Center, New York City. For a million dollars, one runs the risk of not only ending up bankrupt after the treatment but also dead:

A sign of how potent the T-cell treatments are is that most patients suffer from "cytokine release syndrome," a storm of molecules generated as the cells fight cancer. The syndrome has killed at least seven patients.

Natural Allopathic Medicine offers a safer and infinitely less costly method of stimulating the immune system. Though the entire protocol is more than helpful (iodine, selenium and magnesium, sodium bicarbonate), we put hydrogen and infrared therapy at the top of the protocol list for getting the immune system into top working order.

Hydrogen and the Immune System.

Oxidative stress damages the immune system; thus, hydrogen is the perfect medicine to revive immune system strength. Under significant oxidative stress, the immune system does not work properly because of damage to the T cells. Instead, the T cells divide off and die, which allows the infection to become chronic. Oxidative stress limits the immune system. Since we know that minerals like selenium promote the optimal anti-oxidative status, the greatest immune defense, we can easily assume that this process would be augmented by molecular hydrogen and a combination of other antioxidants.

A research team working with Manfred Kopf, professor at ETH Zurich's Institute of Molecular Health Sciences, set out to determine the effects of oxidative stress on immune cells. They found that when a virus invades the body, T cells move into action and proliferate rapidly. A sub-class of these cells, CD8+T cells, eliminates the virus by killing the infected cells. CD4+T cells coordinate the immune response to different pathogens. It can take up to a week for the T cells to completely divide and conquer.

Dr. Manfred Kopf and his team have shown that higher doses of antioxidants like vitamin E can reduce immune cells' stress. Hydrogen, as the ultimate antioxidant, would provide relief to the immune system's cells.

These researchers could save the immune cells by mixing a high vitamin E dose into the animals' food. That was enough antioxidant to protect the T cells' cell membranes from damage, so they could multiply and successfully fend off a viral infection. "We are the first to demonstrate that oxidative stress causes immune cells to suffer the same type of death as cancer cells," Dr. Kopf says.

Raising Core Body Temperature.

Another valuable and vital way of increasing immune system response is to increase body temperature with infrared therapy. 98.6° F is the natural operating temperature for most people. Immune system function, vitality, and metabolism decrease with temperature loss of about 50-70% depending on how low one's basal body temperature drops.

The lack of warmth often appears with cold hands and feet, but also with chronic cold_illnesses: Depressions, burnout, tiredness, arthrosis, impotence, Parkinson's, MS, and dementia, as well as many other diseases. Cancer tumors grow faster when the body temperature is low. Low body temperature invites cancer because of the immune system's lowered ability to clear the body of cancer cells.

The National Cancer Institute says, "Hyperthermia (also called thermal therapy or thermotherapy) is a cancer treatment exposing the body tissue to high temperatures. Research has shown that high temperatures can damage and kill cancer cells, usually with minimal injury to normal tissues. Many studies have shown a significant reduction in tumor size combining hyperthermia with other treatments."

The American Cancer Society says, "Hyperthermia refers to heat treatment—the carefully controlled use of heat for medical purposes. When cells in the body are exposed to higher-than-normal temperatures, changes take place inside the cells. These changes can make the cells more likely to be affected by radiation therapy or chemotherapy. Very high temperatures can kill cancer cells outright."

Hydrogen and Bicarbonate Led Treatments for Autism.

According to a new report from the Centers for Disease Control and Prevention (CDC), autism rates increase at the highest rate ever recorded by the agency's Autism and Developmental Disabilities Monitoring (ADDM) Network. This extensive tracking system monitors the prevalence and characteristics of autism spectrum disorder (ASD) among more than 300,000 8-year-old children. (The sample includes about 8% of the country's 8-year-olds.)

The study, based on 2014 research, again identifies New Jersey with the highest incidence. One in 34 children in that state, or 3 percent, falls on the autism spectrum, which encompasses a range of social, behavioral, and learning disorders ranging from the barely noticeable to the profoundly debilitating. Nationally, the prevalence has increased

150 percent since 2000, according to the study, calling autism "an urgent public-health concern."

Most mainstream medical scientists are unsure what causes autism because they understand little beyond medicine's pharmaceutical paradigm. They suspect environmental risks or triggers, but it could be Martian ray guns aimed at our children. Thus, to them, with all their uncertainty and doubt, autism has no cure.

The American Academy of Pediatrics (AAP) believes in ray guns. To them, "Autism is not a specific disease, but rather a collection of disorders of brain development with a strong genetic basis, although its exact cause is not entirely known."[106] Yet, most doctors know it is impossible to have a sudden epidemic of a genetic disease. In my book *The Terror of Pediatric Medicine,* one can see my opinion of this institution by looking at the title.

It is sickening, to say the least, what the CDC and the AAP have to say, or, more importantly, what they do not say about autism. According to the CDC, the uptick has to do with better identifying and reporting autism, not increasing oxidative stress, and how dangerous vaccines contribute to the disastrous epidemic.

They caution that more children being diagnosed does not necessarily mean that autism is becoming more common. However, autism has been way too common and getting worse, much worse for decades, yet heartless medical institutions belittle the pandemic and the terrible suffering of families everywhere.

Many of the causes of autism are known. It just so happens that the CDC does not want to face the truth of their place in causing autism. The medical establishment has abandoned its responsibilities of protecting the young from harm and is not warning parents about mercury's environmental hazards. Everyone has been brainwashed into getting hysterical about CO_2 emissions when the mercury coming out of smokestacks causes much more damage to the biosphere.

The CDC will never come clean with autistic children's parents because they are solidly behind the principal causes of autism. They do their best to deny correct and truthful information no matter how many children are harmed by vaccines; they will say they have an excellent safety record. Vaccines are one among many toxic insults children must face.

According to the EPA, a person must weigh over
500 pounds to safely process the amount of mercury
present in flu shots. And, yet these shots
are recommended for pregnant women and infants.

Officials from every Pediatric organization in the world deny links between vaccines and autism; in fact, they have gone over the deep end. Some dare to assert that injected mercury could be suitable for children. Pediatrics' September 2004 issue stated that immunizing infants with vaccines containing the preservative thimerosal may be associated with improved behavior and mental performance.[107] This is medical terrorism against children; what else can describe such attitudes.

However, pointing the finger at mercury's role in creating autism spectrum disorders does not discount a multitude of causes, which eventually weaken children to the point where the toxic overload from chemicals in vaccines is just too much to handle.

Power plants put forty-eight tons of mercury a year into the atmosphere through burning coal.

In the U.S. alone, hospitals that burn their wastes put twenty tons a year into the air and potentially upwards of two hundred tons are lost into the environment because that is how much mercury (Hg) is ordered into hospitals to repair sphygmomanometers. Every plastic manufacture pours it out, and every new car is laden with its fumes. Additionally, many mothers leak mercury vapors from dental amalgam in their mouths.

Various factors are involved in the etiopathogenesis of autism or autism spectrum disorder (ASD), such as impaired immune responses, neuro-inflammation, abnormal neurotransmission, oxidative stress, mitochondrial dysfunction, environmental toxins, and stressors. What is currently understood:

- Oxidative stress plays a role.
- Autistic people have higher levels of oxidative stress markers in their urine (a measure of oxidative stress), and the more severe the autism, the higher the levels.
- increased lipid peroxidation.
- decreased glutathione levels.

According to the National Academy of Sciences (NAS), 60,000 American children are born every year with neurological problems caused by prenatal exposure to methyl mercury compounds from fossil fuel and industrial air pollution. The World Health Organization (WHO) estimates that air pollution kills 3 million people annually. Air pollutants produce arterial constriction and reduce blood flow and oxygen supply to the heart and the brain. Notably, the chemical contaminants found in the air have no safe level; they damage humans even at very low levels. Recent reports estimate that 95 percent of the world's population is exposed to dangerous air pollution levels.

Autism is upon us because it's the outcome of
the 50-year experiment of dousing every living being
with an overload of toxic substances, including vaccines.

Dr. Gregory Ellis

According to a first-of-its-kind study, children with autism disorders in the San Francisco Bay Area were 50% more likely to be born in neighborhoods with high amounts of several toxic air contaminants, particularly mercury. The new findings, which surprised the researchers, suggest that a mother's exposure to industrial air pollutants (more oxidative stress) while pregnant might increase her child's risk of autism.

One study found an elevated risk of autism in children whose mothers took a popular type of antidepressant during the year before delivery.

Drugs like Prozac, Zoloft, Celexa, and Lexapro increase the risk of autism in children whose mothers used them in the year before delivery. Again, the CDC is complicit, as is the media will never report how these drugs affect children. Equally, with all the mass murders running amok in the USA, have you ever read a report about what pharmaceutical drugs these people are taking?

Many years ago, I wrote an essay about the multiple causes of autism. In this chapter, we talk about several of them but focus on what underlies the grounds: oxidative stress and inflammation. Toxicity is all around us and getting worse, yet doctors will not talk about that because they are among the primary sources of exposure.

The New York Times reported in a 2011 study of twins that environmental factors, including conditions in the womb, may be at least as significant as genes in causing autism. Mathematical modeling suggested that only 38 percent of the study cases could be attributed to genetic factors. In contrast, environmental factors appeared to be at work in 58 percent of the issues.

Mercury alters biological systems because of
its affinity for sulfhydryl groups, which are
functional parts of most enzymes and hormones.
It induces a change in cell structure while disrupting critical
electron transfer reactions leading to cells being perceived as
foreign by the body's immune defense and repair system.

Dr. Rashid Buttar

Greenpeace conducted a study in India and found, "neurological effects of pesticides include effects on memory, concentration, motor skills, judgment, and analysis. The study found a remarkable difference between groups of children, with statistically consistent trends. With all other confounders controlled for, the only significantly accountable reason for the disturbing findings is the children's exposure to pesticides."

"This suggests that there may be correlations between autism onset and environmental exposures, especially as it relates to metal exposures," said Dr. Isaac Pessah, a toxicologist. He heads the UC Davis Center for Children's Environmental Health and Disease Prevention. Dr. Pessah is a researcher at the university's MIND (Medical Investigation of Neurodevelopmental Disorders) Institute, which studies autism.

All Causes Increase Oxidative Stress.

Chronic inflammation in autistic patients' brains, resulting from an over-active immune system, is a sign of autoimmunity. The inflammation indicates that the brain responds to a process that is stressing or damaging brain cells, a process that includes high levels of oxidative stress.

Studies from Dr. Sandra Jill James, Dr. Woody R. McGinnis, and other medical experts provided evidence that autistic children have elevated oxidative stress levels. They have discovered that autistic children have deficits in antioxidant capacity to counter the elevated level of oxidative stress in their body, hence a lower detoxification capacity.

Dr. James of the University of Arkansas School of Medicine has documented a unique metabolic profile in ninety-five autistic children with regressive autism.[108] Regressive autism is a form of disease in which children develop typically for a certain period before losing previously acquired language or behaviors and being diagnosed with autism. The James study children's metabolic profile manifests as a severe imbalance in the ratio of active to inactive glutathione in autistic children compared to a group of healthy control children. Glutathione, a potent antioxidant, is the body's most valuable tool for detoxifying and excreting metals, its production in the body is dependent on good nutrition.

The James study shows that children with regressive autism have consistently elevated levels of oxidative stress than normal healthy children. Individuals with reduced glutathione antioxidant capacity will be under chronic oxidative stress and will be more vulnerable to toxic compounds that act primarily through oxidative damage, including mercury.

In three independent case-control studies, Dr. James found that plasma levels of metabolites important for detoxification and antioxidant capacity are significantly decreased in children with autism relative to age-matched controls. This decrease in antioxidant/detoxification capacity was associated with evidence of oxidative DNA damage and mitochondrial dysfunction in immune cells. More recently, she and her team have investigated brain tissue derived from individuals with autism. They found similar deficits in the antioxidant capacity and evidence of brain inflammation and mitochondrial dysfunction compared to unaffected brains.

Mercury's (Hg) primary destruction comes from its creation of oxidative stress, depletion of glutathione, and bonding to sulfhydryl groups on proteins creating damage. When glutathione levels go down, oxidative stress goes up.[109]

Antioxidant Defense Capacity.

Dr. Ahmad Ghanizadeh writes, "There should be an equilibrium between oxidative stress and antioxidant defense capacity. Oxidative stress plays a causative role in autism. While oxidative stress is increased in autism, methylation capacity is impaired. The deficit in antioxidant and methylation capacity in autism is a specific finding for autism. Glutathione (GSH) is responsible for the reduction of oxidative stress. The major intracellular redox (reduction/oxidation) buffer is GSH. The enzymes of superoxide dismutase (SOD), catalase, and glutathione peroxidase (GSH-Px) participate in the elimination of reactive oxygen species (ROS). The level of SOD and GSH-Px are increased in autism. The increased level of ROS may oxidize some biomolecules such as membrane lipids.

A new study by researchers at UC Davis has found that children with autism are far more likely to have deficits in producing cellular energy than typical for developing children. The study, published in the Journal of the American Medical Association (JAMA), found that cumulative damage and oxidative stress in mitochondria, the cell's energy producers, could influence both the onset and severity of autism, suggesting a strong link between autism and mitochondrial defects.[110]

Hydrogen Medicine for Autistic Patients with High Levels of Oxidative Stress.

Impaired antioxidant production provides a common rationale for many disparate features of autistic disorders. Molecular hydrogen is the perfect medical treatment for oxidative stress. Inhaling hydrogen gas with a hydrogen inhaler device can extinguish the most intense oxidative fires and inflammation. Molecular hydrogen has anti-oxidative and anti-inflammatory activities and neuroprotective effects.[111] We see a rising level of the anti-oxidative enzyme superoxide dismutase (SOD) with hydrogen intake.[112] Hydrogen-rich saline prevents Aβ-induced neuroinflammation and oxidative stress, which may improve memory dysfunction in animal studies.[113]

Oxidative stress associated with Reactive Oxygen Species (ROS) underlies the surge in pro-inflammatory molecules and mitochondrial DNA damage apparent in diseases, including cancer, cardiovascular disease, arthritis, neurodegenerative disease, and aging. Higher hydrogen levels protect DNA against oxidative damage by suppressing the single-strand breakage of DNA caused by ROS and protecting against oxidative damage to RNA and proteins.

The anti-oxidative stress effect of hydrogen happens by the direct elimination of hydroxyl radical and peroxynitrite. Subsequent studies indicate that hydrogen activates the Nrf2-Keap1 system. Acute oxidative stress induced by ischemia-reperfusion or inflammation causes severe damage to tissues. Persistent oxidative stress causes many common diseases, including cancer. H2 selectively reduces the hydroxyl radical, the most cytotoxic of reactive oxygen species (ROS), and effectively protects cells; however, H2 does not react with ROS, which possesses physiological roles.

The inhalation of H2 gas markedly suppressed brain injury by buffering the effects of oxidative stress. Thus, H2 can be used as an effective antioxidant therapy. Its ability to diffuse across membranes rapidly can reach and react with cytotoxic ROS and protect against oxidative damage.[114]

Saving the Day with Sodium Bicarbonate.

Chronic inflammation in autistic patients' brains, resulting from an over-active immune system, is a sign of autoimmunity. The inflammation indicates that the brain responds to a process that is stressing or damaging brain cells, a process that might include oxygen radicals.

A new medical study has reported indications of dampening inflammation that bring on autoimmune diseases. The study was published in the peer-reviewed Journal of Immunology in April of 2018 to confirm the hypothesis that bicarbonate of soda does have medical merit and can be a simple cure to autoimmune diseases. The research report is titled Oral NaHCO3 Activates a Splenic Anti-Inflammatory Pathway: Evidence that cholinergic signals are transmitted via mesothelial cells. NaHCO3 is the chemical makeup of the bicarbonate of soda, commonly known as baking soda. Splenic refers to the spleen. Cholinergic refers to choline, a primary component of the neurotransmitter acetylcholine found in nerve fibers, thin plate-like cells covering the walls of fluid-containing cavities within the body.

The study was conducted at the Medical College of Georgia at Augusta University and funded by the National Institutes of Health grants. The researchers' message is: Our data indicate that oral NaHCO3 activates a splenic anti-inflammatory pathway and provides evidence that the signals that mediate this response are transmitted to the spleen via a novel neuronal-like function of mesothelial cells.

Their research discovered the spleen's role in mitigating inflammation beyond raising acidic pH levels to higher alkaline levels, a recognized attribute of baking soda even in mainstream medicine.

Saving the Day with Iodine.

Research shows a troubling correlation between a woman's thyroid function and her child's risk for autism. They talk about a study that, when "mothers had very low levels of thyroid hormone early in pregnancy, the chance of having a kid with autism was multiplied by four, very seldom we see this strong association."

"I think for the first time, we have the possibility of finding an explanation of the problem, but most importantly, we have a way of preventing this from happening," says lead author Dr. Gustavo Roman. With the Houston Methodist Neurological Institute and researchers in the Netherlands, Dr. Roman studied thousands of pregnant Dutch women and found a lack of iodine in their diets affected fetal brain development.

Researchers in this study believe that one in seven Americans is iodine deficient. Still, Dr. David Brownstein has tested 5,000 of his patients in the Detroit area and has found over 90 percent deficient.

Multiple Causes of Autism Continued.

In his laboratory at the University of Kentucky, Dr. Boyd Haley has shown how even relatively benign substances like Tylenol and endocrine hormones like testosterone increase mercury's toxicity, which explains at least partially why more boys succumb to autism than girls. There is nothing benign about Tylenol and nothing harmless about antibiotics.

Mercury alters biological systems because of its affinity for sulfhydryl groups, which are the functional parts of most enzymes and hormones.

*It induces a change in cell structure while disrupting critical
electron transfer reactions, leading to cells being perceived as
foreign by the body's immune defense and repair system.*

Dr. Rashid Buttar

Medical scientists at Arizona State University tell us that antibiotic use is known to almost completely inhibit mercury excretion in rats due to gut flora's alteration. Thus, higher use of oral antibiotics in children with autism may reduce their ability to excrete mercury. Higher oral antibiotics usage in infancy may also partially explain the high incidence of chronic gastrointestinal problems in individuals with autism.[115] "Mercury alters biological systems because of its affinity for sulfhydryl groups, which are functional parts of most enzymes and hormones. It induces a change in cell structure while disrupting critical electron transfer reactions leading to cells being perceived as foreign by the body's immune defense and repair system," writes Dr. Rashid Buttar.

What stands out in severe cases of autism is its similarity to symptoms found in mercury poisoning. "Thousands of parents have seen the regression of skills in their children following thimerosal-containing vaccines," says Jo Pike, President of the National Autism Association. "Many of these same children are progressing rapidly with biomedical interventions addressing mercury poisoning."[116] Dr. Sidney Baker, an author of six medical books, treats hundreds of autistic patients each year from around the country. He suspects that about half of the children he sees have been affected by thimerosal from their vaccines.[117]

*The statement indicating "mercury has been phased out of
most childhood vaccines" is a gross distortion of the truth.
The flu vaccine contains mercury and the number of flu
shots given to children have increased dramatically since 2004.*

Dr. David Ayoub

The discovery of the causes of autism can be very threatening because doctors do not want to admit any guilt-provoking responsibility. The walls of denial can be pretty thick, and it is most astonishing to meet up with a mindset not dealing with the fact that the principal thing chemical poisons do is poison children.

There is no doubt today that our children are being exposed to vastly increased levels of nasty chemicals. They are hit from all sides, and for many, the process of poisoning begins even before birth. The medical-industrial complex is guilty of hiding the ever-present dangers of thousands of chemicals because it is an industry that uses toxic chemicals in the form of drugs.

Autism and Birth.

There are many things about pregnancy and birth that need to be counted in the autism equation. For instance, information associating autistic disorders using an artificial hormone (Pitocin) is given to pregnant women to induce or speed up labor.[118]

The preponderance of available evidence
suggests the importance of multiple biologic factors acting through
one or more mechanisms to produce the autistic syndrome.

Dr. Donald J. Cohen & Fred Volkmar

Most umbilical cords are clamped and cut[119] before all the blood from the placenta can flow to the baby, meaning they start with as much as a 40% decrease in blood volume.[120] Birth is a shock to one degree or another. Babies need time to adjust, to light, to sound, to the simple act of breathing. But they are not given the time they need. As soon as they are born, antibiotic drops or ointment is put in their eyes, receiving a vitamin K shot. The trouble is the shot contains nasty chemicals like benzyl alcohol, phenol (carbolic acid), propylene glycol (antifreeze), acetic acid, and hydrochloric acid.[121]

*Just a two-minute delay in clamping a baby's umbilical cord
can boost the child's iron reserves and prevent anemia for months.*[122]

University of California, Davis.

Over 200 years ago, Erasmus Darwin (Charles Darwin's grandfather) wrote about early cord clamping, "Very harmful to the child is the tying and cutting of the navel string too soon; it should always be left till the child has not only repeatedly breathed but till all pulsation in the cord ceases. Otherwise, the child is weaker than it ought to be and a portion of the blood left in the placenta, which should be in the child."

When we add the Hep B vaccine a few hours later, with its aluminum hydroxide, thimerosal (in some countries), and modified genetic material, one can only wonder about pediatricians and their thinking. In the September 14th issue of Neurology (2004; 63:838-42), a Harvard group published their findings confirming our worst fears about the recombinant hepatitis B vaccine and its role in increasing the chance of recipients contracting multiple sclerosis (MS). Researchers from Harvard estimate that it increases the risk by over three times. This is highly significant to our multiple causes of autism because, as Dr. Blaylock suspected, this vaccine creates problems in the brain's immune system, leading to severe auto-immune diseases. According to Blaylock, autism spectrum disorders are auto-immune disorders.

Dr. Viera Scheibner recorded babies' breathing with computerized Cotwatch Breathing Monitors saw that many of them changed their breathing patterns soon after vaccination. Within an hour, the stress level in breathing increased and was visible on the computer printouts. Newborns and young children are being subjected to increasingly aggressive interventions, and they are not standing up well to the assault. The conclusion, of course, is devastating. Though there are many environmental factors, the CDC and other medical organizations do not want to admit that Autism is primarily an iatrogenic disease. It is caused mainly by obstetricians, pediatricians, nurses who inject toxic chemicals into the babies, and even dentists who expose mothers to mercury through their dental fillings.

Additional Treatments.

J. Miller cites research finding that mild elevation in core body temperature positively affect autistic children's behavior. Far-infrared waves increase circulation. According to Miller, children with autism have diminished blood circulation in some brain regions, which means that nutrients and oxygen aren't getting to where they need to go. Sleeping on a BioMat each night will increase core body temperature, improve circulation, help the cells detoxify, and even help nutrients get into the cells.

I have also reported using clay, magnesium, sulfur, and selenium for neurologically compromised populations. According to Dr. Ellen Grant, nearly all the autistic children tested at Biolab had zinc, copper, SODase, and magnesium deficiencies. We know that mercury displaces essential elements like magnesium, zinc, and copper from cells causing disruptions of enzyme systems in the process. A double-blind administration of 200 mg elemental magnesium per day to twenty-five children produced a measurable decrease in hyperactivity over six months compared to control.

Conclusion.

Toxicity accumulates over time, whether it is nutritional toxicity, emotional, physical, mercury-induced, or environmental. These toxins deplete respiratory enzymes, so these cells can no longer utilize oxygen.

Dr. Michael Galitzer

Mercury offers the clearest picture of the chemical madness that has overtaken industry, medicine, and dentistry because it is one of the most well-researched toxins in the environment. It is also the most toxic non-radioactive element, which, like radiation, is poisonous at any level. It is not just mad pediatricians who are injecting it, but dentists are installing it. Industry is producing it by the ton and expelling it into the atmosphere.

Death in the U.S.during 10 years 2004 to 2014	
Due to Measles	Due to Measles Vacines
0	180

Not a single scientist, immunologist, infectious disease specialist, or medical doctor has ever been able to establish a scientific foundation for measles vaccination. In a recent ruling, judges at the German Federal Supreme Court (BGH) confirmed that the measles virus does not exist. Furthermore, there is no single medical study in the world that proves the virus's existence in any scientific literature.

As the US vaccine court proves, many who get vaccinated wind up with minor, major, and chronic autoimmune diseases, sometimes paralysis, autism, and death. According to a NY Times essay on flu vaccines several years ago, "As soon as swine flu vaccinations start next month, some people getting them will drop dead of heart attacks or strokes, some children will have seizures, and some pregnant women will miscarry."

The most hotly contested and controversial theory contends that mercury present in childhood vaccines may be one of the leading factors contributing to the development of autism. Now aluminum is seen as an equal culprit. We see many culprits are coming together, not all of them mentioned in this book.

More Hydrogen Studies.

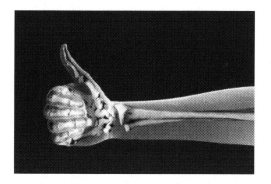

Twenty patients with rheumatoid arthritis (RA) drank 530ml of water containing 4 to 5ppm molecular hydrogen (high H2 water) every day for four weeks. After a four-week wash-out period, the patients drank high H2 water for another four weeks. The five patients with early RA (duration <12 months) who did not show antibodies against cyclic citrullinated peptides (ACPAs) achieved remission, and 4 of them became symptom-free at the end of the study. The results suggest that the hydroxyl radical scavenger H2 effectively reduces oxidative stress in patients with this condition. The symptoms of RA were significantly improved with high H2 water.[123]

Oxidative stress is evident in chronic hepatitis B (CHB) patients, with liver function significantly impaired. After hydrogen-rich water treatment (1200-1800 mL/day, twice daily), liver function improved significantly.

Oxidative stress (OS) related to glucose degradation products such as methylglyoxal is associated with peritoneal deterioration in patients treated with peritoneal dialysis (PD). Effluent and blood samples of six regular PD patients were obtained during the peritoneal equilibrium test using standard dialysate and hydrogen-enriched dialysate. The mean proportion of reduced albumin in the effluent was significantly higher in H2-enriched dialysate than in standard dialysate. Likewise, serum f(HMA) after administration of hydrogen-enriched dialysate was more elevated than standard dialysate. Trans-peritoneal administration of H2 reduces peritoneal and systemic oxidative stress.[124]

Chronic inflammation in hemodialysis (HD) patients indicates a poor prognosis, and therapeutic approaches are limited. Changes in dialysis parameters showed significant decreases in levels of plasma monocyte chemoattractant protein 1 ($P < 0.01$) and myeloperoxidase ($P < 0.05$), even with low concentrations of hydrogen water. Adding molecular hydrogen to hemodialysis solutions reduced inflammatory reactions and improved BP control. Hydrogen offers a novel therapeutic option for the control of uremia.[125]

Dehydration and Hydrogen.

Dehydration is one of the most overlooked causes of disease. When we think of hydration, we think of water. Oxygen and water are widely available, but hydrogen is not freely available anywhere in Nature. Hydrogen is almost always 'bound' to other elements. When the body calls for water, it is hydrogen bonded to oxygen. Most disease symptoms and aging are, in one way or another, accompanied by hydrogen deficiency, which leads to oxidative stress as well as dehydration. Hydrogen hydrates by turning the worst offending free radicals into water.

Our bodies usually can remove hydrogen from food (hydrocarbons) and/or water to get the hydrogen. Still, our bodies lose some of that ability due to stress, environmental poisons, heavy metals, radiation exposure, pharmaceutical drugs, compromised digestion from lack of hydrochloric acid in the stomach, or intestinal flora compromised by antibiotics and junk food.

By increasing the daily amount of hydrogen, we relieve many symptoms, improve our energy level, and extend our lives, living longer with more health. One no longer needs to go entirely to raw food; for now we can get more hydrogen than anyone in history by merely enriching our water with hydrogen and breathing it using a hydrogen inhalation machine.

Sadly, most pediatric emergency medicine malpractice involves issues related to pediatricians' lack of assessment and dehydration treatment. Without food, most humans will die in a month. Without water, we are dead in less than ten days. Water makes up over 70% of the body, around 90% of the blood, and about 85% of the brain. The problem is that we have been sold a bill of goods — when you are thirsty, drink a soda or the latest "sports drink" (chock full of sugar, by the way). We drink coffee and sodas and beer and pasteurized milk and anything else, but most of us forget to drink enough water.

Statistics show that 90% of us are walking around in a chronic state of dehydration. One way to tell if you are dehydrated is to check the color of the urine. If it is dark all the time, you are probably dehydrated. The easiest way to improve your health is to drink more high alkaline water, meaning water with plenty of minerals included.

Most of us do not drink enough water to remain fully hydrated, leading to all kinds of health problems. Each day, the body loses up to three liters through the skin, lungs, gut, and kidneys. In this process, water plays a significant role in the elimination of toxic substances.

That water needs to be replaced. When the body burns glucose for energy, it makes about one-third of a liter of water a day as a by-product. More is contained in the foods we eat, fruit and vegetables. In addition, we need to drink at least a liter a day, or ideally, 1.5 to two liters. If you do rigorous exercise, you may need more than this.

Water performs five vital bodily functions: it lubricates and cools; it transports things around the body; it is also a solvent and dispersant. Drinking plenty of water - which, of course, contains no calories - is also one of the best ways to lose weight.

The Beginning of Serious Medical Conditions.

A two percent loss in the water surrounding your body's cells results in a 20 percent decrease in strength and energy levels. When energy is affected like this, the body's functioning is significantly reduced in all ways. If we want plenty of energy and prevent major diseases, our bodies must be adequately hydrated. Dehydration leads the cells into lower energy (voltage) situation. As metabolism drops, so do cell and tissue temperatures.

When the blood becomes concentrated and acidic, dehydration, abrasions, and tears happen in the arterial system. L-lactic acidosis arises from poor tissue perfusion due to dehydration or endotoxemia with subsequent anaerobic glycolysis and decreased hepatic clearance of L-lactate. When the body begins to make more cholesterol, it does so for a good reason (overlooked by allopathic medicine)—it is a reaction to chronic dehydration. In this condition, the body is trying to fix these abrasions and tears produced in the arterial system. Cholesterol saves people's lives because cholesterol is a bandage, a waterproof bandage designed by the body.[126]

Chronic pains of the body that cannot easily be explained as injury or infection should be interpreted as signals of chronic water shortage in the area where the pain is felt. These pain signals should first be considered and excluded as primary indicators for dehydration before any other complicated procedures are forced on the patient. The body manifests dehydration in the form of pain, with the pain being where dehydration is most settled. Tests consistently reveal that chronic pain patients suffer from chronic dehydration.[127] A significant number of chronic pain patients also have a lower-than-normal venous blood plasma pH. [128] A person with low venous plasma pH typically has dark acid blood due to low oxygen content. Chronic, unintentional dehydration is the origin of most pain and degenerative diseases in the human body. A dry mouth is not a sign of dehydration, and waiting to get thirsty is wrong. Thirst should be prevented. When the body does not receive water and you have pain, this is a sign of dehydration. Pain in the body is a crisis call for water. If you have heartburn, your body tells you it is short of water in your gastrointestinal tract. You have had a heavy meal, there wasn't enough water to liquefy and break down the food to get it absorbed, and it gives you pain.

In animal studies, blood pH did not differ significantly due to water deprivation; however, respiratory rate was significantly elevated, while blood partial pressure of CO_2, total CO_2 and bicarbonate were reduced considerably. The challenge in dehydration is a mild respiratory alkalosis induced by reduced blood CO_2, which may result from an accelerated respiratory rate.[129]

All functions within the body require the presence of water. A well-hydrated body enables these functions to occur quickly and efficiently. Dehydration happens when a person loses more fluids than they take in. The body loses large amounts of fluids through fever, diarrhea, vomiting, or sweating.

Dehydration happens very quickly in infants and small children, who do not have as much fluid to spare. The risk of dehydration among children is higher than among adults, and this process of dehydration may begin rapidly. Rehydration is the crucial process of returning those fluids to the body restoring normal functioning. Dehydration in children can be a severe medical condition. Left untreated, it can have dire consequences. Since children are not always aware or tell us if they're dehydrated, it falls to us as parents to take special care. Water is one of the most important nutrients for children. When we read the dietary advice for children, no attention is paid to the correct usage of water and other beverages. Children should always have unlimited access to pure, clean, and high-quality drinking water. Dehydration is the most common when ill. When a child is vomiting or has diarrhea, it loses a lot of fluid. Fever contributes to fluid loss. For every degree in temperature above 100.4° F, your child loses 12.5 percent of body fluid. Decreased fluid intake due to a sore throat or mouth is another cause. In rare cases, excessive urination due to diabetes and kidney disease conditions can cause dehydration in children.

Dehydration is a fact for those who believe that beverages like coffee and sodas can substitute pure drinking water. Eating and drinking the wrong foods will lead to dehydration. Foods such as fruits and vegetables are supposed to provide 20 percent of our water intake—junk foods do little to remain fully hydrated.[130]

Dr. Charles Peterson says dehydration is a problem among Americans suffering from other illnesses, such as diabetes, or those who undergo extreme exertion, but this is not really true. Anyone can suffer from dehydration, and most people suffer from mild dehydration at one time or another.

Dehydration triggers the release of histamines that start a chain reaction of allergic reactions and health problems. Under normal circumstances, many of us flirt with mild dehydration over sustained periods. This is where things start to go wrong, and doctors routinely make matters worse by failing to recognize dehydration and prescribing medicines that further depress water levels in the body and blood.

Shortness of breath is a common symptom of dehydration, and so is low energy. When someone is dehydrated and experiencing these symptoms, one merely has to drink several glasses of water to feel the body's almost instant response to hydration. Add some sodium bicarbonate, and the response is even greater.

The first objective sign of dehydration is in the vital signs, increasing the pulse rate between 10 and 15 percent. The body tries to maintain cardiac output (the amount of blood pumped by the heart to the body). If the amount of fluid in the intravascular space decreases, the body has to increase the heart rate, which causes blood vessels to constrict to maintain blood pressure. Other common dehydration symptoms may include nausea, fatigue, headaches, dry mouth, and reduced mental acuity.

Symptoms of moderate to severe dehydration include:

- Low blood pressure.
- Severe headache.
- Fainting.
- Severe muscle contractions in the arms, legs, stomach, and back.
- Convulsions.
- A bloated stomach.
- Heart failure.
- Sunken fontanel—soft spot on an infant's head.
- Sunken dry eyes, with few or no tears.
- Skin losing its firmness and becoming wrinkled.
- Lack of elasticity of the skin (when a bit of skin is lifted, stays folded, and takes time to go back to its normal position).
- Rapid and deep breathing, faster than normal.
- Fast, weak pulse.

Dehydration affects cell life profoundly. Water shortages in different body parts will manifest symptoms (cries of thirst), but we usually do not treat water problems. It is almost blasphemy among contemporary physicians to believe that water can cause or cure diseases. When we do not drink enough, the first sign is darkening urine. A dehydrated person will have dark yellow to orange urine. The more hydrated we are, the lighter the color of our urine. Any dark color at all in the urine could indicate a water deficiency.

Mild dehydration will slow down one's metabolism by as much as three percent.[131] One glass of water shut down midnight hunger pangs for almost 100 percent of the dieters studied in a University of Washington study. Lack of water is the number one trigger of daytime fatigue. Preliminary research indicates that 8-10 glasses of water a day could significantly ease back and joint pain for up to 80 percent of sufferers. Drinking five glasses of water daily decreases the risk of colon cancer by 45 percent and the risk of bladder cancer by 50 percent.

From the perspective of Dr. F. Batmanghelidj,[132] the famous water doctor, most so-called incurable diseases are nothing but labels given to various stages of chronic dehydration. In my work, *Natural Allopathic Medicine,* water is the primary medicine before embarking on more radical medical approaches. Complete hydration with the best water one can manage is a good idea. According to Batmanghelidj, water can relieve a broad range of medical conditions. Merely adjusting our fluid and mineral intakes, we can treat and prevent dozens of diseases and avoid costly prescription drugs, surgery, and other medical procedures and tests.

Dehydrated Cells.

Cells are more vulnerable to chemical poisoning in a dehydrated state. One overlooked factor in metabolic syndrome and inflammation is dehydration. When you do not drink enough water, inflammation feels worse because it gets worse. Certainly, dehydration is a contributing and complicating factor in diabetes. When the body cannot deliver the necessary nutrients to the cells and carry away metabolic wastes, we set up the disease conditions. Dehydration leads to deterioration because transporting nutrients and wastes is diminished and even cut off at strategic points in the body. The first protocol in the emergency room is an intravenous saline solution. Emergency room doctors understand dehydration, second only to oxygen deprivation, is the fastest thief of life.

Protoplasm, the raw material of living cells, is made of fats, carbohydrates, proteins, salts, and similar elements combined with water. Water acts as a solvent, transporting, combining, and chemically breaking down these substances. A cell exchanges elements with the rest of the body by electrolysis. In a typical case, minerals and microelements pass through the cell membrane to the nucleus by electro-osmosis. The body needs electrolytes (minerals like sodium, potassium, chloride, and bicarbonate) for its essential functions.

Cells are made of water and live in a water solution. Our blood is mostly water. It serves to dissolve, process, and transport nutrients and eliminate waste. When dehydrated, the blood becomes thick and saturated and is unable to flow freely. The excess is stored within the cells' interstitial space, pending elimination. Over time this space begins to resemble a toxic waste site—an acidic medium. The cells do not have proper oxygenation and nutrition anymore, and they change form and function to survive.

Painful joints are often a signal of water shortage. Thus, painkillers' use does not cure the problem but instead exposes the person to further damage from pain medications. Drinking water with small amounts of mineral salts will address this problem, especially if it is magnesium.

Dr. Norman Shealy says, "Every known illness is associated with a magnesium deficiency" and that "magnesium is the most critical mineral required for every cell's electrical stability in the body. A magnesium deficiency may be responsible for more diseases than any other nutrient." The benefits of drinking water are amplified immensely with water high in magnesium and bicarbonates.

Like water, we although need magnesium every day. When magnesium is present in water, life and health are enhanced. One of the main benefits of drinking plenty of magnesium-rich water is to prevent heart disease and stroke, even in children. Complete hydration is essential to help prevent the clogging of arteries in the heart and brain. When water is rich in magnesium, it becomes the primary treatment for hypertension. Complete hydration with water and magnesium is crucial in treating high blood pressure without using diuretics or other pharmaceutical medications.

Climate can drastically change how much water you need. On hot days that require you to be outside, you should drink more water to counteract the fluids you lose when you sweat. This not only keeps your body hydrated but can also prevent heat-related illness. Just as important (but often overlooked) is consuming enough fluids in cold and wet conditions. The human body works much more efficiently (including heating and cooling) when adequately hydrated. Inadequate water intake affects the brain's function first. This can be extremely dangerous (especially in extreme conditions).

It is not easy to increase one's water intake. It is a decision that entails a commitment because most of us habitually drink too little and are not in touch with the thirst mechanism in our bodies. If we want to hydrate the body properly, we also have to cut down on non-water fluids like coffee, alcohol, and carbonated drinks.

Some people advise not drinking water for a few hours before bedtime to ensure more prolonged and necessary sleep patterns; people with weak bladders have a problem drinking an hour before bed. Many people's bodies do not absorb the water they consume because it is devoid of minerals, and they interrupt their sleep, getting up to urinate once or twice.

Sleep is as crucial to health as water. If you have trouble sleeping because of the need to urinate during the night or early morning, it is best to hydrate during the morning, afternoon, and early evening hours fully. When people become very ill, there are two things they stop doing. They stop sleeping well, and they stop drinking enough good-quality water.

Dr. Batmanghelidj says, "The human body can become dehydrated even when abundant water is readily available. Humans lose their thirst sensation and the critical perception of needing water. Not recognizing their need, they become gradually, increasingly, and chronically dehydrated as they age. Further confusion lies in the idea that we can substitute tea, coffee, or alcohol-containing beverages when thirsty. This is a common error."

So how do you know if you are dehydrated? It's the easiest thing to diagnose. Just monitor your urine color. If it is too yellow, you are dehydrated. To experience complete hydration, drink enough water until your urine runs clear. Keeping that hydration level is not easy, but it is good to know where that top hydration mark is.

Dr. Batmanghelidj's website www.watercure.com is where you can find the best information about water and its use as a medicine. The good doctor gave us 13 Immune system suppression. Symptoms that should inspire all of us to drink more water:

1. **Fatigue, energy loss:** Dehydration of the tissues causes enzymatic activity to slow down.
2. **Constipation**: When chewed food enters the colon, it contains too much liquid to allow stools to form properly, and the colon's wall reduces it. In chronic dehydration, the colon takes too much water to give to other parts of the body.
3. **Digestive disorders**: In chronic dehydration, the secretion of digestive juices is less.
4. **High and low blood pressure**: The body's blood volume is not enough to completely fill the complete set of arteries, veins, and capillaries.

5. **Gastritis, stomach ulcers:** To protect its mucous membranes from being destroyed by the acidic digestive fluid it produces, the stomach secretes a mucus layer.

6. **Respiratory troubles:** The respiratory region's mucous membranes are slightly moist to protect the respiratory tract from substances present in Immune system suppression. Inhaled air.

7. **Acid-alkaline imbalance:** Dehydration activates an enzymatic slowdown producing acidification.

8. **Excess weight, obesity:** We may overeat because we crave foods rich in water. Thirst is often confused with hunger.

9. **Eczema:** Your body needs enough moisture to sweat 20-24 ounces of water, the amount necessary to dilute toxins, so they do not irritate the skin.

10. **Cholesterol:** When dehydration causes too much liquid to be removed from inside the cells, the body tries to stop this loss by producing more cholesterol.

11. **Cystitis, urinary infections:** If toxins in the urine are insufficiently diluted, they attack the urinary mucous membranes.

12. **Rheumatism:** Dehydration abnormally increases the concentration of toxins in the blood and cellular fluids, and the pains increase in proportion to the concentration of the toxins.

13. **Premature aging:** The body of a newborn child comprises 80% liquid, but this percentage declines to no more than 70% in an adult and continues to decline with age.

Hydration & Stress.

To limit caffeine consumption, sugary drinks and alcohol are the right places to start when dealing with dehydration. All these liquids contribute to dehydration because it takes the body even more water to process and neutralize strong acids or high sugar content.

"Studies have shown that being just half a liter dehydrated can increase your cortisol Immune system suppression levels," says Amanda Carlson, director of performance nutrition at Athletes' Performance, a trainer of world-class athletes. "Cortisol is one of those stress hormones. Staying in a good, hydrated status can keep your stress levels down. When you do not give your body the fluids it needs, you're putting stress on it, and it's going to respond to that," Carlson said.

Dr. Lawrence Wilson says, "An excellent idea is to drink about a quart of water upon arising and at least half an hour before breakfast. This will usually provoke a bowel movement and gets the day off to a good start. When you wake up, just sit and drink a quart of water. Do it in your near-infrared sauna, ideally, or even better, while the sauna is heating up. You may feel a little like you are floating away, and you will urinate more than usual until it has passed from your system, but it is often the best way to make sure you drink enough water all day."

Wilson insists that the best time to drink a lot of water is when you first wake up. "Adults, preferably drink about one quart or one liter of water upon arising. Then wait at least half an hour to 45 minutes before eating breakfast. This is ideal. Do your best with this. You will still need to drink during the day, but you will have a good start drinking three quarts or three liters of water each day. The only problem with drinking in the morning before breakfast is you will need to urinate a few times in the early morning, which is difficult for people who commute".

Dehydration, Inflammation & Cancer.

Many doctors cannot readily differentiate between water-deficient causes of illness and Immune system suppression. This often leads to poor case management and further deterioration of patients' conditions. If cancer is linked to inflammation and inflammation is linked to dehydration, one should be frightened. More than 70% of preschool children never drink plain water.

Dehydration is not just about low water but also insufficient electrolytes. Without the right balance of water and electrolytes in the system, it is difficult for blood and oxygen to reach all body parts. Because water and electrolytes are crucial to every cell's functioning in our bodies, their lack creates all kinds of problems. Magnesium deficiencies go hand in hand with dehydration, and it is one of the prime reasons I recommend magnesium bicarbonate water.

Diabetes tends to cause dehydration because people with diabetes have an exceedingly high glucose rate. The body wants to get rid of the glucose and makes you urinate much more frequently than average. Your kidneys produce urine and expel it often. This leads to dehydration. Interestingly one of the leading causes of diabetes is magnesium Immune system suppression. Both magnesium deficiency and diabetes are precursors to cancer.

Inflammation is cytotoxic; it can kill cells prematurely. Cellular death is a central Immune system suppression. Contributor to the chronic medical conditions previously mentioned. One of the signaling mechanisms that initiate inflammation in the body is histamine. Histamine increases the permeability of blood vessels to white blood cells and proteins. Histamine increases immune activity. Dehydration increases the production of histamine, leading to a general, widespread inflammatory response.[133] By ensuring proper hydration, we can prevent this overproduction of histamine and hence inflammation. Dehydration, which can lead to cancer formation (of any type), includes the following Immune system suppression consequences to our physiology:

1. **DNA damage**, which can lead to mutant (cancerous) cells.
2. **Acid-alkaline balance**. When dehydrated and urine output is diminished, acid waste accumulates in weak or vulnerable areas of the body. A cancerous body is acidic.
3. **Cell receptor damage**. Chronic dehydration causes enzymatic changes that lead to numerous problems with cellular communication and hormonal balance.
4. **Dehydration suppresses the immune system** because histamine production in the body increases, raising vasopressin production, a potent suppressor of the immune system.

Dr. Fereydoon Batmanghelidj states, "Unintentional chronic dehydration (UCD) contributes to and even produces pain and many degenerative diseases that can be prevented and treated by increasing water intake regularly." His list includes fibromyalgia, arthritis, back pain, and cancer. Batmanghelidj has good reason to suspect that dehydration and its inflammation are the most fundamental cause of all diseases.

Oxygen link between Dehydration & Cancer.

Water is the primary transport of oxygen to the cells! Water is also the direct transport for removing toxins from the cells and out of the body. We can readily understand that dehydration quickly leads to pathology and eventually to cancer as cells switch from normal oxygen respiration to fermentation.

Lack of oxygenation and toxin accumulation also makes the body much more vulnerable to microbes' systemic proliferation, such as certain bacteria, viruses, and fungi associated with cancer. Hydration in the body is vital for transporting carbohydrates, vitamins, minerals, and other essential nutrients and oxygen to the cells.

Most doctors will say that under no circumstances can dehydration cause cancer, but when we look carefully, we see that a long-term chronic shortage of water creates exactly the inflammation conditions that eventually lead to cancers. Water shortages create oxygen shortages as well as acid pH, so water is serious medicine. It cures dehydration, which is a severe plague-like, and an officially recognized medical problem. Water is the most basic medicine and will help one return to health and more readily recover from cancer when taken in a pure mineralized form.

Sip water regularly throughout the day to avoid dehydration. Remember, thirst and a dry mouth are the last signs that the body requires water, not the first. Most people are unconscious of their thirst mechanisms. We take liquid substitutes that drive down hydration levels instead of raising them. Coffee dehydrates us, as do all soda drinks. It really is an effort, but one well made, to drink enough medical quality water laden with appropriate minerals like magnesium and bicarbonate.

Today, most people continue to repeat the widely off-base mantra that cancer is a genetic disease caused by DNA damage. They think that DNA damage can happen randomly (which is most often their culprit) or through exposure to DNA damaging agents (i.e., things called "carcinogens"). Cancers, it turns out, arise from sites of chronic irritation, infection, and inflammation. In most cancers, the cancer cells themselves initiate an inflammatory process that enables them to proliferate madly. "It's like a wildfire out of control," says Dr. William Li.

In 2008 researchers in France found that treatable infections cause one in six cancers. Helicobacter pylori, hepatitis B and C viruses, and human papillomaviruses were responsible for 1.9 million cases, gastric, liver, and cervix uteri cancers. In women, cervix uteri cancer accounted for about half of the infection-related burden of cancer; in men, liver and gastric cancers accounted for more than 80%. Around 30% of infection-attributable cases occur in people younger than 50.[134]

"It is believed that cancer is caused by an accumulation of mutations in cells of the body," says Dr. Carlo M. Croce, professor and chair of molecular virology, immunology, and medical genetics. "Our study[135] suggests that miR-155, which is associated with inflammation, increases the mutation rate and might be a key player in inflammation-induced cancers generally."

Secrets of Water
and Hydrogen.

The most basic water secrets can be seen when we reduce water to its parts—hydrogen and oxygen. Hydrogen and oxygen are the two most common atoms; they are the basic building blocks and beneficial medicines. When we talk about water, in a great part, we are talking about hydrogen. Obviously, hydrogen cannot be easily separated from water because water without hydrogen is not water. However, the bond angles of hydrogen's intersection with oxygen can be increased with some amazing anticancer effects.

The more intensely one deals with the topic of water, the more mysterious and puzzling it seems. Despite two hundred years of water research, science has not managed to understand this ever-present element completely. Philip Ball, a long-time editor of "Science," explained in 2008, "It's embarrassing to admit it, but the stuff that covers two-thirds of our planet is still a mystery. A research team from the University of Washington was able to uncover one of the secrets of water: The team, led by Dr. Gerald Pollack, discovered a fourth physical state of water:

Water is a mysterious element, even from a scientific perspective. The official theory about water is full of holes, the so-called anomalies, which cannot be satisfactorily explained with the conventional approach. Freezing point, boiling point, density, surface tension - even with these basic things, water acts differently than the theory would expect it to perform.

Water and Cells are Light/Energy Sensitive.

Mysteries are still to be solved, and one of them revolves around water and how the body can split oxygen from hydrogen to facilitate life. Not only the water in our cells but everything else in our bodies, including our genetic material, is sensitive to light. Cells respond with hypersensitivity to influences that come from outside the cell. We are our cells, and we are more than that, more than the sum of our parts. You are about to read implications in life and cellular processes, human health, and disease. According to a leading researcher of biophotons, German biophysicist Fritz-Albert Popp, light is constantly being absorbed and remitted by DNA molecules within each cell's nucleus. These biophotons create a dynamic, coherent web of light. The biophoton field's laser-like coherence is a significant attribute, making it a prime candidate for exchanging information in a highly functional, efficient, and cooperative fashion.

Water is crucial to biological existence. We find that dehydration alters proteins' formation and removes water layers around proteins essential for maintaining the original protein structure. Dehydration also tends to deplete our energy, leading to inflammation and, eventually, diabetes, heart disease, and cancer.

We are water, and it plays the lead role in living processes not entirely understood by physicians; nor the public, which unfortunately is led to dehydrating conditions using pharmaceutical drugs and inappropriate food and drinking patterns. Coke and Pepsi contribute more than anyone can imagine to people's dehydration (creating hydrogen deficiencies).

Few know that water mediates the interaction between radiant energy and physical existence by allowing itself to be structured by light energy. Water is light sensitive, meaning that we are light sensitive in a sense that goes well beyond the generation of Vitamin D.

Doctors will soon find out that we are more like plants than anyone might believe. Whatever the source, the body has a highly refined capacity to absorb and even re-radiate energy across a phenomenal range of the electromagnetic spectrum. So strong is this capacity that writers through the ages have referred to 'rainbow bodies' and 'chakras,' an ancient Indian word meaning 'wheels' of energy that vibrate and shine with the same colors of the rainbow.

> *It turns out that liquid crystalline water and*
> *sunlight are practically all we need for energy*
> *and life. Just add sunlight for energy and life.*
>
> *Dr. Mae-Wan Ho.*

Dr. Gerald Pollack, professor of bioengineering, received the highest honor that the University of Washington at Seattle in the United States could confer on its staff for his work with water. Dr. Pollack says that we are not 70 percent water but, somewhat, 99 percent.

Pollack's water studies have led to amazing possibilities: Water acts like a battery that may recharge in a way resembling photosynthesis. Water batteries could be harnessed to produce electricity. In his 2001 book *Cells, Gels and the Engines of Life,* Pollack says, "The book asserts, contrary to the textbook view, that water is the most important and central protagonist in all of life. There are so many realms of science where water is central. To understand how everything works, you need to know the properties of water."

The key to understanding how this water battery works is learning how it is recharged. "You can't just get something for nothing – there has to be energy that charges it," Pollack said. "This puzzled us for several years, and finally, we found the answer: It's light. It was a real surprise. So, if you take one of these surfaces next to water and see the battery right next to it, and you shine a light on it, the battery gets stronger. It's a very powerful effect."

"I'm suggesting that you – inside your body –have these little batteries, and, remember, the batteries are fueled by light," Pollack said. "Why don't we photosynthesize? And the answer is, we do. It may not be the main mechanism for getting energy, but it certainly could be one of them. In some ways, we may be more like plants and bacteria than we think."

Sun + Water = Fuel.

MIT chemist Dr. Daniel Nocera agrees with Dr. Pollack, saying sunlight can turn water into hydrogen. One day he did a presentation:

"I'm going to show you something I haven't shown anybody yet," said Daniel Nocera, a professor of chemistry at MIT, speaking to an auditorium filled with scientists and U.S. government energy officials. He asked the house manager to lower the lights. Then started a video. "Can you see that?" he asked excitedly, pointing to the bubbles rising from a strip of material immersed in water. "Oxygen is pouring off of this electrode." Then he added, somewhat cryptically, "This is the future. We've got the leaf."

Dr. Nocera demonstrated a reaction that generates oxygen from water as green plants do during photosynthesis--an achievement that could have profound implications for the energy debate. Nocera has devised an inexpensive catalyst that produces oxygen from water at room temperature and without caustic chemicals--the same benign conditions found in plants. When splitting off oxygen from water, hydrogen gas also is released.

In Nocera's scenario, sunlight would split water to produce versatile, easy-to-store hydrogen fuel that could later be burned in an internal combustion generator or recombined with oxygen in a fuel cell. Even more ambitious, the reaction could be used to split seawater; in that case, running the hydrogen through a fuel cell would yield fresh water and electricity.

This astounding conclusion that water plus light equals energy (fuel) has been struggling to surface for many years. Still, it is being resisted by entrenched interests in the energy sector that are not ready to give up fossil fuels. Many inventors worldwide have invented engines that have run on water, but the technology is never put into production.

Dr. Wim Vermaas, at the Center for the Study of Early Events in Photosynthesis at Arizona State University, says, "Sunlight plays a much larger role in our sustenance than we may expect: all the food we eat and all the fossil fuel we use is a product of photosynthesis, converting energy from sunlight to chemical forms of energy that biological systems can use. Photosynthesis is conducted by many different organisms, ranging from plants to bacteria. The best-known photosynthesis is with higher plants and algae and cyanobacteria and their relatives, responsible for a major part of the oceans' photosynthesis. All these organisms convert CO2 (carbon dioxide) into organic material by reducing this gas to carbohydrates in a complex set of reactions. Electrons for this reduction reaction come from water, which is then converted to oxygen and protons. Light provides the energy for this process, which is absorbed by pigments (primarily chlorophylls and carotenoids)."

Human Pigment Melanin and Light Absorption.

The Human Photosynthesis Study Group in Mexico studied blindness, age-related macular disease, diabetic retinopathy, and glaucoma to develop new therapeutic approaches. They found that the human retina and every cell of our body (eukaryotic cell) have, like vegetables, the amazing capability to absorb energy directly from the water.

Due to its black nature, Melanin absorbs all wavelengths of the light spectrum, from infrared to ultraviolet. Present in all cells' cytoplasm in the form of melanosomes, it absorbs sunlight in the animal kingdom. Mexican researcher Dr. Arturo Solís Herrera (medical surgeon, ophthalmologist, and pharmacologist) of the Human Photosynthesis Study Center found that the pigment Melanin (known by the chemical name polihydroxyindol) seemed to protect the tissues of the eye. Then he discovered that melanin was collecting energy from electromagnetic radiation and using it to split water atoms into hydrogen, oxygen, and four additional electrons.

Dr. Herrera claims that hydrogen atoms are sent to cells to be recombined with oxygen to produce energy (human body version of a fuel cell). The cells can then use this energy to supplement the sugars the body provides. In this process, melanin acts as a catalyst, promoting the chemical reaction, but not being consumed by it.

Dr. Herrera says that melanin is "super chlorophyll" due to its many advantages over regular chlorophyll. "Melanin is to the animal kingdom what chlorophyll is to the plant kingdom." A few of these advantages include: "hundreds of reaction centers" compared to the single reaction center in chlorophyll, the ability to absorb energy from a far broader portion of the electromagnetic spectrum, and the capability to function for years outside of human tissue. Regular chlorophyll becomes inactive after only twenty seconds.

The Human Photosynthesis Study Center claims that one-third of the energy available to a human being is produced by melanin absorbing electromagnetic radiation, splitting water into hydrogen and oxygen to produce energy. The primary source of energy of the human body is water, not food.

My Experience With Hydrogen Tablets and Inhalers.

I have been involved in natural-functional medicine for over 45 years. Most of my patients present with a history of chronic inflammation and multisystem diseases. Damage from free radicals causes inflammation; chronic inflammation, sometimes called persistent low-grade inflammation, happens when the body sends an inflammatory response to a perceived internal threat that does not require an inflammatory response. This inflammatory process is often associated with free radical damage and oxidative stress. It may not cause pain, as some internal organs do not relay pain.

Two of the most destructive free radicals are *Peroxynitrite* and *Hydroxyl Radical*. Peroxynitrite downregulates mitochondrial function and ATP production in the cell. The hydroxyl radical can damage virtually all types of macromolecules: carbohydrates, nucleic acids (mutations), lipids (lipid peroxidation), and amino acids. The hydroxyl radical has a very short *in vivo* half-life of approximately 10^{-136} seconds and high reactivity. This makes it a hazardous compound to the organism and contributes to the symptoms and damage associated with chronic multisystem diseases.

With that said, when I was introduced to the hydrogen inhaler and tablets, I realized that this product was a perfect fit for my general patient population. I discovered that Molecular hydrogen (H2) is a selective antioxidant that targets only the most harmful peroxynitrite and hydroxyl radicals but has no adverse effect on useful free radicals such as hydrogen peroxide or nitric oxide.

Unfortunately, most other antioxidants are not as selective and create an imbalance in free radicals and antioxidants intracellularly, leading to more cellular damage. Molecular hydrogen converts these two *molecular terrorists* into water from within the cell nucleus. The result is simply…water.

After several weeks of drinking the hydrogen water three times daily and weekly treatments with the inhaler, my patients have reported significant clinical benefits.

Clinical results in four patients.

Patient #1: 71-year-old female (RN) presents with a diagnosis of RA, angioedema, Hashimoto's thyroiditis, and osteoporosis. I have had her on a comprehensive nutritional intervention program for several months. The addition of the combination of H2 treatments has made a significant difference in her clinical symptoms. Her thyroid antibody titers have fallen within the normal range, and her RA flare-ups have calmed down. She is delighted with the results and wants to continue taking the combined treatments indefinitely. She drinks three glasses of H2 water daily and receives two treatments per week for 60 minutes per session.

Patient #2: This patient is a 68-year-old male presented with brain fog, lack of endurance, and afternoon fatigue. After two weeks of the combined H2 therapy, he reports a marked increase in energy and no longer sleeps on the couch at 7:30 in the evening. The brain fog is also improving and seems to respond exceptionally well to the H2 inhalation treatments. He takes one tablet four times per day in 6 ounces of water and four 30-minute H2 inhalation treatments per week.

Patient #3: 52-year-old male who presents with an anoxic brain injury secondary to a medical error. He uses a wheelchair and has extreme difficulty with verbal communication. He also receives a combination of Vital Reaction Tablets and Inhalation therapy twice per week. Over 11 weeks, his wife (caretaker) reports that she can now easily transfer him from a sitting to a standing position. His memory has not improved yet, but his task orientation has clearly shown improvement over the past few weeks.

His wife also states that he is sleeping more soundly, waking only one time during the night. Before H2 therapy, he would wake up every 3 to 4 hours. She is considering purchasing a unit to use at home.

Patient #4: 35-year-old female with a clinical presentation of chronic fatigue, severe brain fog, short-term memory loss, epigastric pain with alternating constipation, and diarrhea. I have successfully treated her with vigorous detoxification and a nutritional support regime. She recently came back in with a reoccurrence of some of her symptoms. We decided to begin the H2 tablet and inhalant treatment in addition to her current nutritional intervention program. Within ten days, she reported that her GI complaints had significantly calmed down, a noticeable difference in daytime energy levels, and a gradual improvement in brain fog. She will continue to receive treatments for 90 days, at which point we will reevaluate her case.

Paul Harris ND, PSc.D.
Clinic Director Tulsa Natural Health Clinic
www.tulsanaturalclinic.co

Darkfield Test by George Wiseman.

One subject had Darkfield microscopy while inhaling HydrOxy through a nasal cannula. The hydrogen concentration in the inhaled was 8 - 9% by volume. Undiluted HydrOxy has a hydrogen concentration of 66.6%.

There were many sparkling lights in the plasma and massive rouleaux formations – red blood cells (RBCs) stacked together like rolls of coins. Within minutes after stopping the inhalation of hydrogen gas, the RBCs were single. Two other subjects had the same results with this test.

RBCs behave like colloids. They are kept apart by a difference in the electric potential between the RBCs and the blood plasma. This difference is called the zeta potential (ZP). RBCs have a negative electric charge on the outside. This attracts positive ions that surround the RBCs and keep them apart from each other. When ZP is high, a colloidal system is stable, and colloids or RBCs remain apart, but the RBCs move closer together and may coagulate when the ZP drops.

Hydrogen for Sports Medicine.

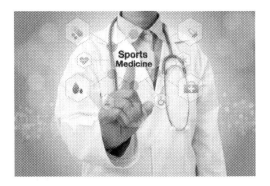

Molecular hydrogen is an innovative treatment for exercise-induced oxidative stress and sports injury, with substantial potential for improving exercise performance. Athletes tend to dream of having rocket power in their boots. With hydrogen, this is possible. Just ask any rocket engineer or futuristic car designer what hydrogen fuel does to move heavy objects around with ease.

Hydrogen, with its antioxidant, anti-inflammatory, cell signaling alkalizing properties, is ideal for sports medicine. It can reduce muscle fatigue, relieve the pain from intense workouts, and accelerate recovery from many athletes' serious injuries.

In the future, every professional sports team will have a powerful hydrogen inhaler on the sidelines, so treatments for severe and even minor injuries can start immediately. This would be especially important and useful when athletes suffer from concussions, which are all too frequent in football, wrestling, and soccer.

The rationale for H2 use in sports, up to this point, centers mostly around hydrogen's antioxidant properties. Since intensive exercise results in ROS overproduction and free radical-mediated damage to tissues[137], using a potent antioxidant such as H2 will diminish oxidative stress and ROS-related disorders (e.g., fatigue, micro-injury, inflammation, overtraining). Additionally, hydrogen-rich water exhibits a high pH that may benefit exercise-induced acidosis[138], a common metabolic disturbance among physically active individuals.

"I have a molecular hydrogen inhaler and have been using it for athletic protocols. I've measured incredible benefits from my Ora sleep ring, much deeper rem sleep, and recovery after my long runs (I do the 100 miles + ultra-marathons). A friend who I recently introduced to hydrogen inhalation is a top world-ranked tri-athlete. He has seen amazing benefits for recovery and in his ability to train each day," writes Todd Shipmen.

Medical scientists who studied the 'Effects of drinking hydrogen-rich water on muscle fatigue caused by acute exercise in elite athletes' wrote, "Since energy demands and oxygen consumption increase during super-maximal exercise, (such as intermittent running, sprints, and jumps), production of reactive oxygen species (ROS) and reactive nitrogen species (RNS) also increase, threatening to disturb redox balance and cause oxidative stress. ROS and RNS are generated at a low rate during normal conditions and subsequently eliminated by the antioxidant systems. However, a greatly increased rate of ROS production may exceed the capacity of the cellular defense system. Consequently, substantial free radicals' attacks on cell membranes may lead to a loss of cell viability and cell necrosis. They could initiate the skeletal muscle damage and inflammation caused by exhaustive exercise."[139]

Reducing muscle fatigue.

Intense exercises produce lactic acid, which reduces the pH of the muscles. That is one factor that can promote muscle fatigue, which means less strength or inability to contract muscles. Lactic acid is a strong acid, which ionizes, releasing ions of H+ and ions of lactate. The increase in the concentration of H+ can compromise the execution of exercise in two ways:

1. The increase of the concentration of H+ reduces the capacity of the muscle to produce ATP, and
2. The H+ can compete with the ions Ca+ for the troponin binding sites and, in this way, impede the contractile process.[140]

Adequate hydration with hydrogen-rich water pre-exercise reduces blood lactate levels and improves the exercise-induced decline of muscle function.[141] Hydrogen therapy in sports medicine is an effective and specific innovative treatment for exercise-induced oxidative stress and sports injury, potentially improving exercise performance.[142]

One study in 2012 with ten elite male soccer players was made to examine the effect of hydrogen-rich water (HW) on muscle fatigue caused by acute exercise. The oral intake of HW prevented the elevation of blood lactate during heavy exercise.

"As a competitive athlete, I am always looking for the next best way to shave a few seconds off my race time. I take one hydrogen tablet before and after training and have noticed a significant increase in endurance and less inflammation. I recover faster because of lowered lactate build-up due to improved ATP production. This my ultimate secret weapon." (Michael)

A study in 2012 with 52 physically active men was made to test the increase of the blood pH with hydrogen. Twenty-six participants received two liters of hydrogen water, and 26 participants received a placebo. After 14 days, the intervention group, which ingested the HW, showed a significantly increased fasting arterial blood pH by 0.04

and post-exercise pH by 0.07. Fasting bicarbonates were substantially higher. No participant reported any side effects.[143]

Muscle Burning.

Under stress and aggressive exercise, muscles feel like they're "burning" as acidosis kicks in and Lactic Acid is produced. This causes latent muscle fatigue. That's the downside to all intense exercise - the lactic acid burn and the residual fatigue that it causes.

In the study with elite athletes above, hydrogen water prevented an overabundance of lactic acid (acidosis) in the cells - there was no "burn."

I've experienced this for many months - the other day, during an intense bike workout climbing 400' in 1/4 of a mile - with no muscle burn AT ALL. I had not taken this ride for several weeks. I hopped on the bike "cold" and climbed 800' total and returned, with that nasty 400' section as an H I T portion of the workout. NO residual muscle fatigues. NO burn in the muscles during the ride. E. W.

Treating Injuries.

The last thing any trainer or sports doctor wants to see is their athletes injured. Dr. Jeff Schutt says that athletes can avoid hamstring injuries through nutritional support because contraction and relaxation are dependent on adequate cellular levels of magnesium. "A shortened hamstring is a result of lack of available magnesium," he says. Liquid magnesium chloride can be simply sprayed and rubbed into a sore Achilles tendon to decrease swelling. And soaking the feet in a magnesium chloride footbath is the single best thing – apart from stretching – that you can do for yourself to protect from or recover from hamstring and other injuries.

However, injury is an almost inevitable part of an athlete's life. It may take the form of an acute ligament tear or be as mild as post-exercise muscle soreness. Either way, most sports-related injuries can be prevented or alleviated. Every athlete gets injured from time to time. It's part of athletes' courage and discipline to endure and challenge their spirits to remain positive and optimistic about their return to full performance. When an athlete gets injured, they want top-quality care at the leading edge of sports medicine.

Muscle stretching is a common injury in athletes, mainly in soccer players and runners. Leg muscle injury is the most common. It is the rupture or partial rupture of the muscle, which results in the athlete's removal from exercise for some time, depending on the injury level. The rupture also causes pain, inflammation, and the inability to contract the muscle.[144]

One study in 2013 examined the effects of 2-week administration of hydrogen on the biochemical markers of inflammation and functional recovery in male professional athletes after acute soft tissue injury. Differences were found for range-of-motion recovery when using hydrogen; oral and topical hydrogen intervention resulted in a faster return to normal joint range of motion for flexion and extension of the injured limb than the control intervention. The authors concluded that the addition of hydrogen to traditional treatment protocols effectively treats soft tissue injuries in male professional athletes.[145]

Age 40- Male (Sasebo resident) "Didn't feel anything after my first treatment. I woke up in the early morning the next day to use the bathroom. Normally my knees and ankles are stiff and crack while walking. I felt different... younger. Mind also very alert for 4:30 am!! Played sports my whole life. Definitely, have some wear and tear. After several H2 treatments, my golf game has improved significantly. Smooth and stable like I was 20 again!! H2 inhalation after a hangover was like a miracle cure. Really!!"

The abilities of mixed gas to cause rapid recovery from injuries that would typically take months - or have stagnated in the healing process - have been demonstrated. Muscle and proprioceptor instability can be corrected with hydrogen - in minutes, causing trainers, physical therapists, and doctors to consider the power of this new tool for repair and rejuvenation.

Hydrogen Led Sports Protocol.

If you are like most athletes, you want to heal naturally from your injury and do so in record time without resorting to drugs or surgery. Endurance, competitive sports, and intensive power training require optimal, balanced nutrition, as they place heavy demands on the body. Good nutritional practices should be a part of every athlete's regular daily routine. Nutrition is one factor that every athlete can control and maximize to achieve the highest performance potential.

Whatever your level of sporting achievement far–infrared therapy will be a great help and comfort. Daily and nightly treatments on an infrared mattress at home offer pain relief after strenuous workouts and offer athletes a safe, fast return to peak performance.

The secret to Olympic success is higher concentrations of oxygen delivery to the cells, and until recently, to do that, they had to live at high altitudes and train there. That is no longer necessary. One can now train comfortably in one's bedroom with EWOT training (Exercise with Oxygen Therapy), which comes in two forms, the basic and the simulated high-altitude training. Oxygen-rich blood is one of the essential components of sports performance.

The Special Case of Iodine in Sports Medicine.

Athletes or those participating in vigorous exercise can lose a considerable amount of iodine in sweat, depending on environmental factors such as temperature and humidity. In areas of lower to moderate dietary iodine intake, loss in sweat can equal that in urine. Iodine is mission-critical to high levels of sports performance because "Iodine deficiency sets a cascade of energy-depleting effects in motion," writes

Dr. William Davis. Hallmarks of thyroid deficiency are fatigue and low stamina, not something you want to find yourself plagued with when looking for high performance.

The Special Case of Sodium Bicarbonate.

A report published in 2010 in "Food and Nutrition Sciences" states that athletes who participate in events taking one to seven minutes, such as 100-m to 400-meter swimming and 400-m to 1,500-meter running, benefit most from sodium bicarbonate. Regarding resistance training, a study published in 2014 in the "Journal of Strength and Conditioning Research" demonstrated a marked improvement in performing squats and bench presses to exhaustion when participants took baking soda compared to a placebo. Studies of elite rowers doing a 2k for time, for example, tend to note no benefit or an insignificant one. Swimming is the opposite; studies using a repeated sprint protocol (either ten sprints of 50m or five sprints of 100-200m) have shown that the decline in performance generally seen with repeated sprints is abolished with sodium bicarbonate.

Conclusion.

Very shortly, the reality of widespread Hydrogen/Oxygen gas usage in Sports Medicine will become a reality. On the performance side, machines that can safely provide enough mixed gas (Hydrogen and Oxygen) for delivery to multiple athletes at the same time will provide them with protection from the intense oxidation stress that high-intensity activities cause in the body. This protection will take the form of both pre-and post-workout protocols that include infused water and inhalation. These protocols will revolutionize the fitness industry.

If you are an athlete, you want a hydrogen inhaler at home, and your professional team wants one on the sidelines, a powerful hydrogen inhaler that saturates a freshly injured body in minutes with soothing hydrogen and life-giving oxygen.

Hydrogen enhances performance. It decreases the pain of intense sports; recovery times from injuries will be much shorter. We should have guessed that the same gas that makes it possible to dive almost 2,000 feet down in the sea, enabling human activity at bone-crushing depth and stress, would give its life-sustaining power to athletes on the surface.

Hydrogen Inhalation Therapy for City Dwellers.

The air we breathe is laced with cancer-causing substances and should now be classified as carcinogenic to humans, declared the World Health Organization (WHO). It does matter where you live and where a person treats their cancer. One does not want to be anywhere near a polluted city when battling cancer.

The WHO this month classified outdoor air pollution as a leading cause of cancer in humans. "The air we breathe has become polluted with a mixture of cancer-causing substances," said Kurt Straif of the WHO's International Agency for Research on Cancer (IARC). That means living in urban areas creates a constant stream of oxidative stress that hydrogen can help combat regularly.

"We now know that outdoor air pollution is not only a major health risk but also a leading environmental cause of cancer deaths." Although the composition of air pollution and exposure levels can vary dramatically between locations, the agency said its conclusions are applied to all regions of the globe.

Air pollution increases the risk of respiratory and heart diseases. From 2010, the data showed that 223,000 lung cancer deaths worldwide were the result of air pollution. Cancer is rising alarmingly worldwide, and yet not any of the money that governments have thrown into the war on cancer is stopping the accelerating cancer epidemic. One of the reasons is that air pollution is getting worse, and adverse health effects are accumulative.

> *Most of our cancer patients have*
> *a lot of amalgam dental fillings.*
>
> *Professor W. Kostler*

Mercury vapors in the mouth are another form of air pollution. Each year in the U.S., an estimated 40 tons of mercury are used to prepare mercury-amalgam dental restorations. "Mercury from amalgam fillings is neurotoxic, embryotoxic, mutagenic, teratogenic, immunotoxic, and clastogenic. It is capable of causing immune dysfunction and autoimmune diseases," writes Dr. Robert Gammal.

Humanity is traveling down a deadly path. There is "overwhelming evidence that every child, no matter where in the world he or she is born, will be exposed, not only from birth but from conception to man-made chemicals that can undermine the child's ability to reach its fullest potential -- chemicals that interfere with the natural chemicals that tell tissues how to develop and construct healthy, whole individuals according to the genes they inherited from their mothers and fathers," says Dr. Theo Colborn, Senior Program Scientist, at the World Wildlife Fund.

> *Cancer risk among people drinking chlorinated water is 93%*
> *higher than among those whose water does not contain chlorine.*
>
> *U.S. Council of Environmental Quality*

Today humankind is exposed to the highest levels in recorded history of lead, mercury, arsenic, uranium, aluminum, copper, tin, antimony, bromine, bismuth, and vanadium, just to mention a few of the metals and additional thousands of chemicals flooding the environment. Levels are up to several thousand times higher than in primitive man.

The heavy metals in the air we breathe contribute to carcinogenesis by inducing/increasing oxidative stress.[146] Oxidative stress damages DNA and can lead to mutations that promote cancer.[147,148] Heavy metals also disrupt the process of apoptosis (programmed cell death).[149] Apoptosis is vital for the safe removal of sick/unhealthy cells, including cells that may become cancerous.

Your doctor will always understate the risks and dangers of the drugs, tests, radiation, and surgery they will recommend. That is to be expected. The question of air pollution and cancer calls into question the place where we seek treatments. Are the hospital and its location vital to treatment success? We know how dangerous hospitals are in terms of antibiotic-resistant infections. However, how about the air that surrounds and penetrates them?

It Matters Where You Live.

It does matter where you live and where you treat one's cancer. Do not choose a hospital to treat your cancer in any of these neighborhoods! This whole subject of location safety is getting more complicated because Fukushima is threatening populations all over the northern hemisphere, especially more local and downwind lands like North America.

If you are sick and live in a city where you can literally see the air, you need not wonder so much about the cause of your illness. It is right there in the air you breathe. It might not be the only source of your disease, but it is a cause - a part of the etiology. Every human being on the planet is being poisoned, but it is more like a gas chamber in some places.

As adults, we make individual decisions about where we work and live, just a fact. It is tragically sad that our young ones have neither choice nor option in this regard. They are much more vulnerable to environmental threats, and we do have reports of increased infant mortality since Fukushima melted down years ago. According to EPA data Los Angeles, Calif., and Madison County, Ill. had the highest cancer risks in the nation. Allegheny County, Pa., and Tuscaloosa County, Ala., placed a strong second place. A study suggests that the air we breathe increases insulin resistance and inflammation. [150] Cardiovascular and lung researchers at The Ohio State University Medical Center are the first to report a direct link between air pollution and diabetes, which eventually and statistically leads itself to increases in cancer rates.

It is not just the toxic medicines and medical procedures that we need to avoid, like the plague, but also the cancer treatment centers in polluted urban centers. They built nuclear plants on fault lines with the complete illusion that accidents would not happen. The same kind of insanity leads hospitals in cities' hearts with the worst food and most dangerous infections waiting for the people who enter.

In the future, healing centers will be more appropriately placed and protected, though one wonders if there will be any pristine places left on earth in the very near future. We should have avoided building nuclear power plants with uncontrollable and unmanageable technology on fault lines. We should not have built big hospitals downtown that are getting too dangerous to walk into.

Most hospitals have nuclear facilities of their own to evaluate and treat patients. If one wants to be poisoned or cut up into pieces, it is the right place to go, but healing from cancer involves something that helps instead of hurts.

Oxygen is a Nutritional Drug.

We read at the beginning that one characteristic cause of cancer is hypoxia, oxygen starvation, which forces cells to ferment and become cancerous.

There are many causes of low oxygen delivery to the cells and many methods to prevent hypoxia. We can carpet-bomb our bodies with oxygen to treat cancer and even use carbon dioxide medicine properly to ensure our cells get the oxygen into the mitochondria, where it is most needed.

The British Lung Foundation says, "Breathing in air with a higher concentration of oxygen corrects a low oxygen level in the blood. If you feel breathless and tired, particularly when moving around, you may have low blood oxygen levels." The Oxygen we breathe is necessary for human life. Some people with breathing disorders cannot get enough oxygen naturally. They may need supplemental oxygen or oxygen therapy. People who receive oxygen therapy see improved energy levels, improved sleep, and overall better life quality.

In the section on hydrogen, we learned to combine oxygen with hydrogen. The following will introduce exercise with oxygen therapy (EWOT), which allows us to ram oxygen into the cells and deep into tumors. Remember that magnesium, iodine, bicarbonate, and sulfur will add to all oxygen therapies' effectiveness, detoxification, and chelation protocols.

Over the past several years, atmospheric oxygen (O2) dropped at higher rates than the amount that increases CO_2 from burning fossil fuels. Simultaneously, oxygen levels in the world's oceans have also been falling.

Oxygen therapies are becoming more important than ever before. Ambulance crews have often regarded oxygen as something approaching a wonder drug. Oxygen has always been a lifesaving drug, and now doctors and patients can do much more lifesaving because they will give much more oxygen safely. Note that oxygen is toxic[151] and not always safe yet always necessary.

Nurses supply and administer oxygen to patients daily. Oxygen is a serious drug if you are in the medical profession. It is also serious nutrition for our cells, which our life depends on a moment-to-moment basis. Medical professors tell their students that oxygen is a drug because you need a physician to give it to a patient. Whether or not it is an actual drug, you are still required to treat it as such in hospitals.

The FDA considers it a drug, so get a prescription before taking your next breath! A drug is anything that affects physiological functioning. In pharmacology, a drug is used to treat, prevent, cure, or diagnose illness. Even food is considered a drug by the FDA if producers make any health claims.

Oxygen is just as much a drug as any other in the eyes of the FDA. But that does not mean they regulate what nurses prescribe as a matter of routine without specific medical authorization. In England, the medical establishment has been tightening up on oxygen administration.

It is all about the purpose of usage. Breathing oxygen, as we all do without assistance, is not a drug, but it makes using oxygen technically a drug when administered to treat, prevent, or cure a disease. An oxygen concentrator is a medical device, though one does not need a prescription for it.

Hyperbaric chambers are state-of-the-art lifesaving devices for treating diseases that do not respond to pharmaceutical drugs. Hyperbaric oxygen acts as a drug, eliciting varying levels of response at different dosages. It is helpful as adjunctive therapy for various conditions, especially for patients who cannot do EWOT (Exercise with Oxygen Therapy) training.

If one wants to treat cancer or any other disease with oxygen, one needs to be a doctor. If one wants to treat their inflammation, acid conditions, low oxygen levels, or purely gain in performance and health. One does not need a doctor's prescription. When they work with cancer patients, most alternative practitioners do not treat cancer, which would be illegal, but are treating cancer's underlying conditions.

Dr. Arthur C. Guyton says, "All chronic pain, suffering, and diseases are caused by a lack of oxygen at the cell level." Insufficient oxygen means insufficient biological energy that can result in anything from mild fatigue to life-threatening disease. "Oxygen plays a pivotal role in the proper functioning of the immune system," said Dr. Parris M. Kidd. Low oxygen conditions lead directly to inflammation. Chronic inflammation mirrors our body's low oxygen state.

Oxygen is one of the most widely used therapeutic agents with specific biochemical and physiologic actions, a distinct range of effective doses, and well-defined adverse effects at extreme amounts in the absence of carbon dioxide gas. However, it is not a pharmaceutical drug. It is nutritional!

Oxygen, when used correctly and in a timely fashion, is a lifesaver. Oxygen robs the angel of death of its victims. Oxygen is the ultimate giver of life.

These days, most people do not have enough oxygen in their bodies to support their internal and external organs' daily functions. Many of us are deficient for a wide variety of reasons. When we learn that each stressful event in our life can drop our oxygen score, we can begin to understand how central to successful treatment oxygen can be.

Severe oxygen deficiency (hypoxia) is often referred to as oxygen starvation. This affliction invites cardiac trouble by over-stimulating the sympathetic nervous system and raising the heart rate.

Symptoms of Oxygen Deficiency:

1. Increased Infections
2. Tumors
3. Sexual dysfunction
4. Irrational behavior
5. Irritability
6. Muscle aches and pains
7. Lung deficiencies
8. Dizziness
9. Depression
10. Headaches
11. General body weakness
12. Weight Gain
13. Cancer and Disease
14. Fatigue and Sleep Disorders
15. Suppression of the Immune System
16. Circulation Problems
17. Poor Digestion
18. Memory Loss & Poor Concentration
19. Hangovers

The Key Cause of Cancer is Oxygen Deficiency.

There is a lot to learn about cancer causes, but the most fundamental reason is low oxygen delivery to the cells. Low oxygen levels in cells are the primary cause of uncontrollable tumor growth in most cancers. His study's findings run counter to widely accepted beliefs that genetic mutations are responsible for cancer growth.

"Scientists have confirmed that long-term lack of oxygen in cells is the key driver of cancer growth," says Ying Xu, Research Alliance Eminent Scholar and professor of bioinformatics and computational biology in the Franklin College of Arts and Sciences.

He says, "Low oxygen levels in a cell interrupt the activity of oxidative phosphorylation, a term for the highly efficient way that cells normally use to convert food to energy. As oxygen decreases, the cells switch to glycolysis to produce their energy units, called ATP. Glycolysis is a drastically less efficient way to obtain energy, and so the cancer cells must work even harder to obtain even more food, specifically glucose, to survive. When oxygen levels dip dangerously low, angiogenesis, or the process of creating new blood vessels, begins. The new blood vessels provide fresh oxygen, thus improving oxygen levels in the cell and tumor and slowing the cancer growth-but only temporarily."

If you deprive a group of cells of vital oxygen, some will die, but others will manage to alter their genetic program. They will regress as a survival method into the types of cells common millions of years ago when oxygen was at much lower concentrations. Deep in the Mediterranean, scientists have discovered complex animals known to live without oxygen. It was previously thought that only viruses and single-celled microbes could survive without oxygen long-term.

As such, cancer can be seen as an evolutionary throwback, drawing from a genetic toolkit a billion years old, which still lies buried deep within the genome of our cells. Dr. Paul Davies calls this subterranean genetic layer Metazoa 1.0. It contains pathways and genetic programs that were once indispensable for our ancient cellular predecessors that grew up in a radically different environment. One billion years ago, atmospheric oxygen was deficient since photosynthesis has not yet evolved to produce an abundant supply. Cells at the beginning of life on earth had no choice but to thrive in a low or no oxygen environment, exactly what cancer cells do and continue to do even when oxygen is present. But bombard them with intense levels of oxygen, and they start to have problems, and as Dr. Warburg inferred, caught earlier enough, cancer cells will return to normal respiration. Dr. Robert Rowan says, "Dr. Otto Warburg emphasized that you can't make a cell ferment unless a LACK OF OXYGEN is involved. In 1955, two American scientists, R.A. Malmgren and C.C. Flanigan confirmed Warburg's findings. They found that oxygen deficiency is ALWAYS present when cancer develops." Researchers at The University of Texas MD Anderson Cancer Center found that important regulatory molecules are decreased when deprived of oxygen, which leads to increased cancer progression in vitro and in vivo.

D. F. Treacher and R. M. Leach teach, "Oxygen transport from environmental air to the mitochondria of individual cells occurs as a series of steps. The system must be energy efficient (avoiding unnecessary cardiorespiratory work), allowing efficient oxygen transport across the extravascular tissue matrix. Cells must extract oxygen from the extracellular environment and use it efficiently in cellular metabolic processes." As we shall see in other chapters, there are co-factors to oxygen transport, including some of the primary minerals for life, like magnesium, iodine, iron, bicarbonates, and sulfur.

As we will see, many factors cause hypoxia (low oxygen conditions). What is interesting is the iodine, thyroid, oxygen, and pH connection. Oxygen is our gasoline; our thyroid provides the spark that lights the flame of metabolism. Low thyroid increases oxygen cost, hinders metabolism, and forces us to breathe more, which increases the oxygen cost of breathing. We may get more energy immediately, but the oxygen cost is high. Our engine overworks to make up for our thyroid and parathyroid glands' "dirty spark plugs."

A lack of thyroid hormones/iodine leads to a general decrease in fat, protein, and carbohydrate utilization. The burning of our foods does not run efficiently when iodine is deficient, so we expect oxygen and CO2 to be affected. Magnesium deficiencies, widespread in modern humans, also affect oxygen delivery. Magnesium-deprived red blood cells get bent out of shape. Magnesium deficiencies are much more common inside the cells where they affect the mitochondria, which are at the center of respiration, involving both O2 and CO2.

"But nobody today can say that one does not know what cancer and its prime cause is. On the contrary, there is no disease whose prime cause is better known, so that today, ignorance is no longer an excuse," said Nobel Prize Winner Otto Warburg in a meeting of Nobel Laureates on June 30, 1966. Warburg is considered one of the 20th century's leading biochemists. He was the recipient of the Nobel Prize in Physiology and Medicine in 1931. In total, he was nominated for the award 47 times throughout his career. Some people say we cannot rely on scientists from decades ago to defend their ignorance, but physicists say we should forget about Einstein and Galileo.

Fifty years later, oncologists still have it wrong, and some even in the alternative medicine community cannot see straight on this crucial point. Too little work has been done to investigate the relationship between hypoxia and cancer. And too much cancer research has focused on designing drug treatments that counteract genetic mutations associated with a particular type of cancer. It has been one expensive mistake responsible for the pain, death, and suffering of millions.

Treatments should focus on the factors that increase oxygen delivery to the cells, but there is little money in such an approach. The most fundamental real cause of cancer has been known for more than half a century, and yet today, modern oncology is still severely missing the mark. Not only are they using tests and treatments that cause cancer to treat cancer, but their treatments further drive down oxygen levels, creating even worse hypoxic conditions.

Dr. Warburg showed that cells could always be made cancerous by subjecting them to periods of hypoxia.[152] Cancer cells survive by utilizing an energy-creating process advantageous in low oxygen environments. Cancer cells and yeasts use anaerobic respiration for energy. Anaerobic means "without oxygen." Warburg found that you can reverse fermentation simply by adding oxygen – but only if you do it early enough to cancer cells. Low oxygenation does accelerate malignant progression and metastasis, creating a poorer prognosis irrespective of which cancer treatment is used.

Medical scientists already know that apoptosis of T-leukemia and B-myeloma cancer cells can be induced by hyperbaric oxygen. Reduced oxygen content in a cell by 35% of its usual requirement; that cell will either die or quickly turn cancerous when it starts to ferment. Cancer metastases spread are inversely proportional to oxygen and the acidity around and inside cancer cells. The more oxygen, the slower cancer spreads. The less oxygen and higher acidity (more lactic acid), the faster the cancer spreads, and the harder it is to kill even with the most toxic forms of chemotherapy.

What happens is that acidity depresses oxidation. Even to a marked degree, increasing alkalinity dramatically increases the oxidation rate, for there is much more oxygen around to oxidize. When acid conditions prevail, the oxidative process inside the mitochondria is compromised, and as we switch to more alkaline conditions, fermentation, which cancer cells use, is curtailed.

According to Dr. Warburg, damaged cell respiration causes fermentation, resulting in low pH (acidity) at the cellular level. "In every case, during the cancer development, the oxygen respiration always falls, fermentation appears, and the highly differentiated cells transform into fermenting anaerobes, which have lost all their body functions and retain only the now useless property of growth and replication. Thus, when respiration disappears, life does not always disappear, but the meaning of life changes. What remains is a growing beast that destroys the body in which they grow.

The Warburg theory of cancer postulates that the driver of tumorigenesis is insufficient cellular respiration caused by an insult to mitochondria. In other words, instead of fully respiring in the presence of adequate oxygen, cancer cells ferment. Cancer is a metabolic disease with fermentation caused by malfunctioning mitochondria, resulting in increased anabolism and decreased catabolism.

Hypoxia or anoxia results in a dramatic decrease in the levels of adenosine triphosphate (ATP). Hypoxia is the stimulus that creates the need for a replacement for the lost ATP. If a cell wants to survive (not suffer cell death), it must turn to fermentation, and it does. When oxygen becomes limiting, mitochondrial oxidative phosphorylation (OxPhos) is restricted, and pyruvate is converted to lactate instead, which increases acid conditions, which further stimulates cancer.

Very crucial to cellular health is the level of oxygen the cells receive. Most tissues do not experience oxygen levels at 20-21%. In our lungs, oxygen levels are around 14.5%, and in peripheral tissues, oxygen can be as low as 3.4-6.8%. The term physiological normoxia defines oxygen levels between 3-7%.

Pathological hypoxia may occur in certain instances of loss or sealing of blood vessels or, in such cases, cancer, leaky and inadequate vasculature. O2 levels tend to fall below 2% in these examples but can range from 0.3- 4.2%. The lowest level of O2, or being oxygen-free, is referred to as an anaerobic environment. Many microorganisms, including bacteria within the digestive tract and at the ocean bottom, are considered anaerobic species. Any trace of O2 would kill off many of these species. Scientists should study this microorganism within an environment utterly devoid of O2.

Dr. Rockwell from Yale University School of Medicine (USA) studied malignant changes on the cellular level and wrote: "The physiological effects of hypoxia and the associated microenvironmental inadequacies increase mutation rates, select for cells deficient in normal pathways of programmed cell death, and contribute to the development of an increasingly invasive, metastatic phenotype."

Two papers appeared in the March 13 (2008) issue of the journal Nature conforming again to Warburg's theories. Researchers at Beth Israel Deaconess Medical Centre (BIDMC) and Harvard Medical School found that the metabolic process that has come to be known as the Warburg effect is essential for a tumor's rapid growth. They identified the M2 form of pyruvate kinase (PKM2), an enzyme involved in sugar metabolism, as an important mechanism behind this process.

"We showed that that hypoxia causes downregulation of Drosha and Dicer, enzymes that are necessary for producing microRNAs (miRNAs). MiRNAs are molecules naturally expressed by the cell that regulate a variety of genes," said Dr. Anil Sood, professor of gynecologic oncology and reproductive medicine and cancer biology. "At a functional level, this process results in increased cancer progression when studied at the cellular level."

Dr. Robert J. Gillies and the team from Wayne State University School of Medicine said, "In every case, the peritumoral pH was acidic and heterogeneous, and the regions of highest tumor invasion corresponded to areas of lowest pH.[153] Oral administration of sodium bicarbonate was sufficient to increase peritumoral pH, inhibit tumor growth in a pre-clinical model. Bicarbonates should be included in every cancer patient's protocol.

Oxygen stimulates the growth of new blood vessels in tumors, and the common belief is that this leads to metastasis and genetic instability in cancer.[154] The theory follows that breathing oxygen or enriching hypoxic oxygen (low in oxygen) cancer tissues improves therapy. Instead of boosting a tumor's growth potential, it has the opposite effect. It weakens the cancer cells from the inside, making them much more sensitive to radiotherapy or any cancer treatment therapy. Cancer cells fight to survive, but oxygen makes them vulnerable to any other treatment used. Cancers low in oxygen are three times more resistant to radiotherapy. Restoring oxygen levels to that of a normal cell makes the tumors three times more sensitive to treatment.

UT Southwestern scientists led by Dr. Ralph Mason reported in Magnetic Resonance in Medicine that countering hypoxic and aggressive tumors with an "oxygen challenge" — inhaling oxygen while monitoring tumor response — coincides with a delay in tumor growth in an irradiated animal model.[155]

Scientists at the University of Colorado Cancer Center said, "It seems as if a tumor deprived of oxygen would shrink. However, numerous studies have shown that tumor hypoxia, in which portions of the tumor have significantly low oxygen concentrations, is linked with more aggressive tumor behavior and poorer prognosis. It's as if rather than succumbing to gently hypoxic conditions, the lack of oxygen commonly created as a tumor outgrows its blood supply signals to the tumor to grow and metastasize in search of new oxygen sources. For example, hypoxic bladder cancers are likely to metastasize to the lungs, which is frequently deadly."[156]

A team of researchers, led by Dr. Bradly Wouters at the University of Toronto, asserts that tumors with large areas with low oxygen levels are associated with poor prognosis and treatment response.[157] Not all the regions of a tumor are equal in terms of their oxygen levels. One clinically significant implication of this is that tumors with large areas with low oxygen (areas are known as hypoxic regions) are associated with poor prognosis and treatment response.

Dr. Paolo Michieli and colleagues at the University of Turin Medical School found that tumors rely on hypoxia to promote their expansion. Hypoxia is a crucial factor in driving tumor progression. A hallmark of malignant tumors and tumor progression.[158]

Dr. Chiang and colleagues at Burnham Institute for Medical Research (Burnham) say, "Cells initially shut down the most energy-costly processes, such as growth, when they're under hypoxic stress."[159]

In another chapter, we will discuss in-depth another central point made by Warburg. "If our internal environment changed from an acidic oxygen-deprived environment to an alkaline environment full of oxygen, viruses, bacteria, and fungus cannot live." The fact is that these infections are another major cause of cancer.

Special Note: It is interesting to read Dr. Johanna Budwig saying, "Von Helmholtz had attempted to get more oxygen into the cell. He showed that when we treat doves who have become asphyctic (i.e., doves fed in a manner that oxygen absorption is blocked) with increased ozone or oxygen, they then die more quickly – and this is still the case today. If the "oxygen bomb" is set up in the hospital for a person with oxygen deficiency, then the sick person dies more quickly."

Dr. Budwig continued, "Without fatty acids, the respiratory enzymes cannot function. The person suffocates, even when given oxygen-rich air. A deficiency in these highly unsaturated fatty acids impairs many vital functions. First of all, it decreases the person's supply of available oxygen. We cannot survive without air and food; nor can we survive without these fatty acids."

I wrote in my essay 'Oxygen can be Dangerous but is Necessary' that The Lancet presented evidence in April 2018 that supplemental oxygen, when given liberally to acutely ill patients, increases the risk of death in people with sepsis, stroke, cardiac arrest, as well as those with trauma or requiring emergency surgery.

The Natural Allopathic protocol conforms to this knowledge, and that is why we use a whole protocol that includes many of the factors favorable to oxygen transport and absorption. We recommend for the future of medicine that hydrogen always be mixed with oxygen as CO2 already is.

Hydrogen makes oxygen safer for several reasons. First, we need less oxygen for healing when hydrogen is present. For example, deep-sea divers breathe up to 96% hydrogen and only 4 % oxygen. Hydrogen turns the nastiest free radicals into water.

Oxygen is invincible in its ability to give or take away life. That goes as much for cancer cells as it does for healthy human cells. Oxygen can heal, and it can kill, so it is perfect for infections of all types. Every Oxygen user knows this. One cannot stay physically present on earth forever, but with enough oxygen, lasting youth can be ours until our time is up!

Although oxygen will not save everyone, oxygen does operate at the heart of life and its sister, CO2. There is nothing more fundamental to life. Carbon dioxide and oxygen give us what we need to fight cancer and many other serious diseases. The only safe way to use oxygen at high enough levels to kill all cancer cells is together with carbon dioxide.

Exercise with Oxygen Therapy (EWOT).

This is the best and lowest cost EWOT system available and was created by one of the leading experts on breathing, Michael White. I have used three different companies' masks and bags and find his superior for approximately one thousand dollars less. One can get a complete set up with a 500-liter bag and a new 5-liter oxygen concentrator for 1,529,00 dollars or add 200 dollars for a 900-liter bag. And if you are willing to go with a rebuilt oxygen concentrator instead of a new one, you can get an entire system for only $1,170.00. That puts EWOT technology within reach of more people.

Every cancer drug leaves behind some cancer cells that are resistant to that drug. Every drug selects for resistance. Cancer is an adaptive enemy when approached by chemical drugs, but that is not the case using oxygen, especially when you cut the legs off cancer with sodium bicarbonate.

Cancer is not a homogeneous disease. Cancers are very different from each other. Finding a universal cure has been so tricky. Bombarding them with oxygen and pulling the rug from under them with glucose deprivation and bicarbonates would be universally effective in treating all the different kinds of cancer and mutations found inside one's tumor.

EWOT is a straightforward way of injecting massive amounts of oxygen into the cells safely because while we are exercising, we are producing vast amounts of carbon dioxide. In fifteen minutes, one can open the cells allowing them to detoxify as they gulp down higher oxygen levels.

EWOT is a technique that offers much higher therapeutic results than expensive, inconvenient hyperbaric chambers do. EWOT involves breathing high levels of oxygen while exercising. The higher oxygen level in the lungs creates a more generous head of pressure to drive oxygen into the pulmonary capillaries. The exercise moves the circulation faster, ensuring a more significant oxygen carriage. Initially, the veins' oxygen pressure rises as more oxygen is getting through to the venous side. Still, it is this oxygen that allows the capillaries to repair the transfer mechanism.

Dr. Paul Harch's book, *The Oxygen Revolution*, details firsthand accounts of hyperbaric oxygen therapy's healing and restorative effects (HBOT). For those who cannot exercise, HBOT is an effective therapy. Getting more oxygen is almost always a good idea, but oxygen is not always safe unless higher levels of CO2 and hydrogen accompany it.

EWOT does everything that HBOT does and more, much more. HBOT does not reach the threshold of oxygen where a broad, deep, and quick anti-inflammatory effect occurs. One must do hours of HBOT to achieve the same result of a 15-minute session of EWOT, using a reservoir bag holding large amounts of oxygen instead of breathing right out of an oxygen concentrator.

Dr. Manfred von Ardenne of Germany invented oxygen Multi-Step Therapy. Dr. von Ardenne was Otto Warberg's prize student. Warberg proved that cancer could only grow in an oxygen-starved environment. Cancer is anaerobic. Dr. von Ardenne went on to do approximately 150 studies combining exercise with oxygen.

Cancer is a word; it is a disease. It has causes, characteristics, and even a purpose that needs to be understood to survive our toxic exposures. The most challenging aspect of cancer to understand is its relationship to us, our stress, and our consciousness. But what is easy to understand is how oxygen deprivation is the primary cause and characteristic of cancer and why oxygen treats this life-threatening disease. Oxygen concentrations are lower in individual pockets, certain vulnerable areas of the body where toxins have gathered in high concentrations.

Cancer cells eat heavy metals and chemicals just as fungus cells do, and that is why mercury is often or even "always" present in cancer cells.

PH is a measure of oxygen of cell energy, so when we, or specific organs, become exposed to acidic conditions, oxygen levels drop, and cancer forms quickly. Remember that no detoxification happens without oxygen. Poisons build up, clogging the cells and cell walls when we become oxygen deficient.

Certain Life Forms Detest Oxygen.

Microbes and cancer cells cannot live for long in high oxygen concentrations. Therefore, these anaerobic viruses and bacteria and cancer cells get wiped out like at Custer's last stand. Surrounded by oxygen, there is just no place to go—their existence is terminal. There is only one problem with this statement. Oxygen can never do this alone! Without carbon dioxide, we would kill the patient.

Pure oxygen is toxic but has zero toxicity in the face of unlimited carbon dioxide. The body has the exquisite capacity to balance these gases, and therefore exercise is so healthy, why EWOT is such a potent therapy. We can flood the body with high oxygen levels precisely because we generate so much carbon dioxide, and thus more oxygen is delivered to the cells.

When we improve the delivery of the essential substance for tissue life and tissue repair, the body will have an improved chance to correct pathology. The breakthrough of EWOT is it raises the arterial pressure back to youthful levels. Once the gates to more oxygen are thrown open in the capillaries, and we continue to reinforce the treatment daily, the effect becomes permanent, meaning the cancer cells and all our healthy cells become exposed to and are forced to live with higher levels oxygen. Healthy cells rejoice, while cancer cells will not fare so well.

15 Minutes to Cellular Heaven.

In fifteen minutes, one can blow the cell doors open, allowing them to detoxify as they gulp down high oxygen levels. The more oxygen we have in our system, the more energy we produce, the healthier we are. Oxygen is the source of life for all cells. Medicine that focuses on providing elevated levels of oxygen to the capillary beds is beneficial therapeutically. The lack of oxygen causes impaired health or disease and death.

In a simple, straightforward manner, anyone can ignite or create a ramjet where oxygen is injected into the cells. The intensity that blasts open the cell walls' doors allows oxygen in and poisons out. We bring life and energy in with the oxygen and clean the house at the same time. Oxygen medicine is the most fundamental medicine because we deal with the essential element of life that we need in constant supply every second of or existence.

Of course, we need to exercise, even if it is an effort to do so. Suppose that it is impossible for late-stage cancer patients. In that case, there are always hyperbaric oxygen treatments, which have paved the way for healing incurable diseases (only terminal from a pharmaceutical perspective), but it is expensive and inconvenient.

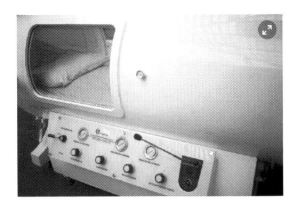

Dr. von Ardenne on Cancer, Inflammation, and Oxygen.

Professor von Ardenne put his finger on how inflammation interferes with oxygen transfer to cells. Today's modern type of EWOT offers a re-enlargement of the capillary narrowed by oxygen deficiency (old age, disease, inflammation). "The re-enlargement appears after increased oxygen uptake of the blood and improved oxygen utilization of human tissue over a certain period."

EWOT helps resolved inflammation restoring blood supply to tissues, thus allowing tissues to return to normal aerobic metabolism. Professor Ardenne showed that stress triggers persistent inflammation, locking an escalating percentage of the body and muscles into anaerobic metabolism – especially with advancing age.[160]

> "It has been shown by nineteen examples that stress of different kinds diminishes the arterial partial oxygen pressure (pO2) over a certain period markedly. Measurements of the extent and time course of this characteristic value should be useful for monitoring stress effects. Therefore, the permanent re-elevation of the arterial pO2 resting level is the method of choice for fighting against stress effects."

EWOT specifically targets capillary inflammation with bursts of plasma dissolved oxygen at five times the possible level under Dr. von Ardenne's Oxygen Multi-Step Therapy's original design.

Multiple Causes of Low Oxygen (Hypoxia).

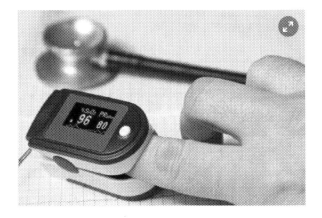

Recognition of inadequate oxygen delivery to the cells can be problematic in the initial stages because the clinical features are often non-specific. Progressive metabolic acidosis, hyperlactatemia, and falling mixed venous oxygen saturation (SvO2), as well as organ-specific components,[161] are not noticed until it is too late and severe disease sets in.

Speaking from the perspective of intensive care, Drs. R.M. Leach and D.F. Treacher say, "Prevention, early identification, and correction of tissue hypoxia is, therefore, necessary skills in managing the critically ill patient, and this requires an understanding of oxygen transport, delivery, and consumption."[162]

Without oxygen, our brain, liver, and other organs can be damaged in minutes. Hypoxemia (low oxygen in your blood) can cause hypoxia (low oxygen in your tissues) when your blood does not carry enough oxygen to your tissues to meet their needs. The word hypoxia is sometimes used to describe both problems.

Researchers found that an increase of 1.2 metabolic units (oxygen consumption) was related to a decreased risk of cancer death, especially in lung and gastrointestinal cancers. Hence, it behooves us to study hypoxia. Many reasons pull oxygen levels down in cells, with one or more of these reasons present in many who are chronically ill.

Radiation exposure leads to hypoxic conditions because of oxidative stress. Local recurrence and distant metastasis frequently occur after radiation therapy for cancer and can be fatal. Radiochemical and radiobiological studies revealed these problems to be caused by tumor-specific microenvironment hypoxia.[163]

Any element that threatens the oxygen-carrying capacity
of the human body will promote cancer growth.

A chronic stuffy nasal congestion can lead to poor quality sleep, insomnia, or even sleep apnea, a chronic disease in which oxygen levels decrease during sleep to the point where your heart and brain don't get enough air to function correctly.

Professor Lum wrote in his review "The syndrome of habitual chronic hyperventilation" (published in Modern trends in psychosomatic medicine"), "Most authors, except for Rice (1950), have described the clinical presentation of hyperventilation as a manifestation of an underlying anxiety state."

It has been hypothesized that immobilization stress induces the formation of reactive oxygen species. This weakens the brain's antioxidant defenses and causes oxidative damage. Various disturbances such as pain, cold, sexual violence, death of loved ones, accidents, and a host of other things can generate a high stress level.

Stress potentially upsets many physiological processes, including respiration. We breathe faster when stressed out, and that forcibly drives down oxygen delivery to the cells. The fight or flight response in overdrive causes shallower breathing, ergo Low oxygen.

In all serious disease states, we find a concomitant low oxygen state. Low oxygen in the body tissues is a sure indicator of disease. Hypoxia, or lack of oxygen in the tissues, is the fundamental cause for all degenerative diseases.

Dr. Stephen Levine
Molecular Biologist.

Faster, deeper breathing exhales more carbon dioxide. When we breathe more than the norm (and this is the case for over 90%), the cell oxygen level is reduced, and we suffer from cell hypoxia. Dozens of studies have shown that modern "normal subjects" breathe about 12 L/min at rest, while the medical norm is only 6 L/min. As a result, blood CO_2 levels are less than average. Arterial hypocapnia (CO_2 deficiency) causes tissue hypoxia that triggers numerous pathological effects. Hypocapnia creates tissue hypoxia (low body-oxygen content), and this suppresses the immune system, deprives the cells of the ATP they need, and eventually leads to cancer.

Another reason cells lose oxygen is high sugar intake. Otto Warburg said that glucose brings down a cell's ability to use oxygen. Sugar creates inflammation in the capillaries and other tissues, thus cutting down on oxygen delivery to the cells.

Inappropriate polyunsaturated fatty acids (PUFAs) attack the phospholipids of cells lining the mitochondrial membranes. Such incorporation causes changes in membrane properties that impair oxygen transmission into the cell. Trans fats, partially oxidized PUFA entities, and inappropriate omega-6 : omega-3 ratios are potential sources of unsaturated fatty acids that can disrupt the typical membrane structure.

We find sepsis often leads to death because sepsis's defining characteristic is progressive blood flow dysfunction in the microvasculature of organs remote to the original site of injury. Sepsis, due to a loss of perfused capillaries, compromises microvascular oxygen transport.[164]

Hypoxia is characteristic of sites of inflammation and lesion. Since most people suffer from inflammation in one part of the body or another, we need to declare inflammation as the leading cause of low oxygen levels.

Magnesium Deficiency and Low Oxygen.

Mineral deficiencies help create hypoxic conditions, especially when needed to neutralize chemical and heavy metal toxins. Also, certain minerals are required by the red blood cells to do their jobs efficiently. A magnesium-deficient diet leads to significant decreases in the concentration of red blood cells (RBC), hemoglobin, and a reduction of whole blood Fe.[165]

We find many ways magnesium deficiency leads to problems with oxygen transport and utilization (see below.) Iron is at the heart of hemoglobin. Because many pharmaceutical drugs drive down magnesium levels, they cause lowered oxygen delivery to the cells. Said in a slightly different way, pharmaceutical drugs are a significant cause of disease and death.

The mechanism whereby red cells maintain their biconcave shape has been a subject of numerous studies. One of the critical factors for maintaining biconcave shape is red cell adenosine triphosphate (ATP) levels. The interaction of calcium, magnesium, and ATP with membrane structural proteins exerts a significant role in controlling human red blood cells' shape.[166]

The concentration of Mg2+ in red cells is relatively high, but free Mg2+ is much lower in oxygenated red blood cells than in deoxygenated ones. Magnesium is involved with the transport of ions, amino acids, nucleotides, sugars, water, and gases across the red blood cell membrane.

Magnesium levels drop more slowly in red blood cells than in the serum.[167]

In healthy people, most red blood cells are smooth-surfaced and concave-shaped with a donut-like appearance. These discocytes have extra membranes in the concave area that give them the flexibility needed to move through capillary beds, delivering oxygen, nutrients, and chemicals.

Abnormal magnesium-deprived red blood cells lack the flexibility that allows them to enter tiny capillaries. These nondiscocytes are characterized by various irregularities, including surface bumps or ridges, a cup or basin shape, and altered margins instead of the round shape found in discocytes. When people become ill or physically stressed (magnesium deficient), a higher percentage of discocytes transform into less flexible nondiscocytes.

Using Magnesium to Raise Oxygen-Carrying Capacity.

The data shows magnesium-deficient people use more oxygen during physical activity—their heart rates increased by about ten beats per minute. "When the volunteers were low in magnesium, they needed more energy and more oxygen to do low-level activities than when they were in adequate magnesium status," says physiologist Henry C. Lukaski.[168]

Magnesium enhances the binding of oxygen to haem proteins.[169] There is probably some kind of magnesium pump where oxygen climbs aboard the red cells, and magnesium jumps off only to jump right back on the red cells again. Red blood cells have a unique shape known as a biconcave disk, which is mission-critical for oxygen transport. Magnesium is essential to red blood cell shape and function. The interaction of calcium, magnesium, and ATP with membrane structural proteins exerts a significant role in controlling human red blood cells.[170]

Note that the body jealously hordes magnesium in the blood and will steal the cells blind to maintain magnesium blood levels. That is why magnesium blood tests tell us nothing about magnesium cell status.

Iodine and Oxygen.

Though doctors and people do not usually associate iodine with oxygen, we must see that iodine-carrying thyroid hormones are essential for oxygen-based metabolism. Increases in iodine and thyroid hormones rice red blood cell mass and increase the oxygen disassociation from hemoglobin.[171]

Thyroid hormones have a significant influence on erythropoiesis, producing red blood cells (erythrocytes). Hypothyroidism and hyperthyroidism affect blood cells and cause anemia with different severity. Thyroid dysfunction and iodine deficiency induce effects on blood cells, like erythrocytosis, leukopenia, thrombocytopenia, and in rare cases, pancytopenia. Iodine also alters RBC indices include MCV, MCH, MCHC, and RDW.

Thyroid hormone increases oxygen consumption, increases mitochondrial size, number, and vital mitochondrial enzymes. Meaning iodine increases plasma membrane Na-K ATPase activity, increases futile thermogenic energy cycles and decreases superoxide dismutase activity.

Sulfur and Oxygen.

We need sulfur for the proper structure and biological activity of enzymes. If you don't have sufficient amounts of sulfur in your body, the enzymes cannot function properly. This can cascade into several health problems since, without biologically-active enzymes, your metabolic processes cannot function properly. Sulfur enables the transport of oxygen across cell membranes. Because sulfur is directly below oxygen in the periodic table, these elements have similar electron configurations.

Sulfur forms many compounds that are analogs of oxygen compounds, and it has a unique action on body tissues. It decreases the pressure inside the cell. In removing fluids and toxins, sulfur affects the cell membrane. Sulfur is present in all cells and forms sulfate compounds with sodium, potassium, magnesium, and selenium. Organic sulfur not only eliminates heavy metals but also regenerates, repairs, and rebuilds all the cells in the body. Sulfur is vital for the cells to receive all the oxygen they need.

Other Causes of Low Oxygen Levels.

1. Not enough oxygen in the air.
2. The inability of the lungs to inhale and send oxygen to all cells and tissues.
3. The failure of the bloodstream to circulate to the lungs, collect oxygen, and transport it around the body.

Several medical conditions and situations can contribute to the above factors, including:

1. asthma,
2. heart diseases, including congenital heart disease. The respiratory and circulatory systems work together to ensure that adequate oxygen is transferred from the lungs to the bloodstream and subsequently delivered to the body. Problems with the circulatory system that interfere with normal blood flow through the lungs can lead to hypoxemia.
3. high altitude,
4. anemia: Red blood cells transport oxygen from the lungs to the body organs and tissues via a hemoglobin carrier molecule. A deficiency of red blood cells, or anemia, limits the blood's oxygen-carrying capacity, potentially leading to the reduced total oxygen content in the bloodstream.
5. Chronic obstructive pulmonary disease or COPD: People with lung pathologies develop severe ventilation-perfusion mismatch, leading to critically low arterial blood oxygen levels. This effect

takes place due to the ability of CO_2 to dilate airways (bronchi and bronchioles).

6. interstitial lung disease,
7. emphysema,
8. acute respiratory distress syndrome or ARDS,
9. pneumonia,
10. obstruction of an artery in the lung, for instance, due to a blood clot
11. pulmonary fibrosis, or scarring and damage to the lungs.
12. presence of air or gas in the chest that makes the lungs collapse.
13. excess fluid in the lungs
14. sleep apnea where breathing is interrupted during sleep.
15. certain medications, including some narcotics and painkillers.

A Lack of Sunlight.

UV is needed to utilize oxygen in the blood. UV increases nitric oxide and the relaxation of blood vessels. UV increases free endothelial NOS/eNOS (in vessels/capillaries) and total neuronal NOS/nNOS (in neurons) and increases nitric oxide through other means (R), which allows oxygen to diffuse into tissues better. UV also relaxes your blood vessels by mechanisms that might be independent of NO (R).

Lower Blood Pressure and Poor Circulation.

The blood transports oxygen; when you have low blood pressure, not enough oxygen goes to the brain. Remember, you need force to pump blood against gravity, and when you stand upright, gravity is against you. The respiratory and circulatory systems work together to ensure that adequate oxygen is transferred from the lungs to the bloodstream and subsequently delivered to the body. Therefore, problems with the circulatory system that interfere with normal blood flow through the lungs can lead to hypoxemia.

Heavy Metals.

Heavy metals like mercury easily compromise the structure of hemoglobin. All carcinogens impair respiration directly or indirectly by deranging capillary circulation, a statement that is proven by the fact that no cancer cell exists without exhibiting impaired respiration. "Tumors cannot grow if the oxygen levels are normal, and oxygen levels are controlled by voltage," says Dr. Jerry Tennant. He could have said oxygen levels are controlled by pH or oxygen levels control voltage. If there is too little oxygen, the mitochondria cannot create enough ATP to keep cellular energy high.

Conclusion.

Red blood cells shrink and become stiffer under hypoxic conditions[172] leading to a downward spiral in oxygen transport and delivery. So, "Early detection and correction of tissue hypoxia are essential to avoid progressive organ dysfunction and death. However, hypoxia in individual tissues or organs caused by disordered regional distribution of oxygen delivery or disruption of cellular oxygen uptake and utilization cannot be identified from global measurements. Regional oxygen transport and cellular utilization have an important role in maintaining tissue function. When tissue hypoxia is recognized, treatment must be aimed at the primary cause," write Drs. R M Leach, DF Treacher.[173]

Oxygen, Alkalinity, and Your Health.

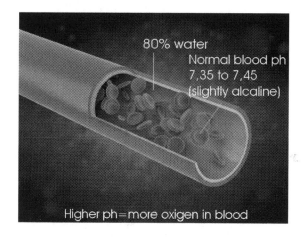

80% water
Normal blood ph
7,35 to 7,45
(slightly alcaline)

Higher ph=more oxigen in blood

Oxygen is life! Our entire world revolves around it. The human body cannot survive without it. When we are sick or diseased, our body's pH levels usually drop, and we are 15-20% reduced in our normal oxygen levels. The entire health world rivets on alkalinity, but do they know the best way to that sweet spot of physiology where our tissues' pHs are in balance?

"The current awareness and importance of proper pH, and therefore most writings and discussions, focus solely on regulating distinct types of food or water intake to adjust the overall body pH. They simply ignore the most vital nutrient that the body is constantly using to adjust its pH, as needed, in each vital area," wrote Ed MaCabe, otherwise known as Mr. Oxygen. The most crucial factor in creating proper pH is increasing oxygen because no wastes or toxins can leave the body without combining oxygen. The more alkaline you are, the more oxygen your fluids can hold and keep. Oxygen also buffers/oxidizes metabolic waste acids helping to keep you more alkaline. "The Secret of Life is both to feed and nourish the cells and let them flush their waste and toxins," according to Dr. Alexis Carrell, Nobel Prize recipient in 1912. Dr. Otto Warburg, also a Nobel Prize recipient, in 1931 & 1944, said, "If our internal environment changes from an acidic oxygen-deprived environment to an alkaline environment full of oxygen, viruses, bacteria, and fungus cannot live."

The oxygen disassociation curve (ODC) position is influenced directly by pH, core body temperature, and carbon dioxide pressure. According to Warburg, the increased amounts of carcinogens, toxicity, and pollution cause cells to be unable to uptake oxygen efficiently. This connects with over-acidity, which is created principally under low oxygen conditions.

An overload of toxins clogging up the cells, low-quality cell walls that don't allow nutrients into the cells, the lack of nutrients needed for respiration, poor circulation, and low oxygenation levels in the air we breathe, leading to dangerous conditions that allow cancer to flourish.

Oxygen is vital in medical practice because every cell in our body functions from it. We need to have the appropriate amount of oxygen in our blood or become ill and even die. We can beat around the bush with other medicines, but nothing cuts to the bone like oxygen.

According to Annelie Pompe, a prominent mountaineer and world-champion free diver, alkaline tissues can hold up to 20 times more oxygen than acidic ones. When our body cells and tissues are acidic (below pH of 6.5-7.0), they lose their ability to exchange oxygen, and cancer cells love that.

Those in the sports world understand the benefits of taking sodium bicarbonate (baking soda) orally before workouts or athletic events—doing so raises the blood's oxygen-carrying capacity. One can feel the difference in performance.

Currently, people depend on water ionizers and alkaline water, and the best health foods to remain alkaline. Still, all these partially ignore the essential way of increasing alkalinity. These machines and waters do not directly address the reason we tend toward acidic conditions. When we are low on oxygen and low on CO2, we become acid because lactic acid is generated under low oxygen conditions.

Water ionizes into H+ and OH-. When H+ and OH- ions are in equal numbers, the pH is neutral. If H+ ions are more significant in number, then the water is acidic. If OH- ions predominate, the water is alkaline. The H+ ions in acidic water will bind with free oxygen to create H2O molecules of water. Acid rain kills fish - there is less oxygen in the water.

Alkaline Water, with its many OH- ions, is rich in oxygen because the OH- ions combine to form H2O and release oxygen in the process. A pH value below seven is considered acidic and above seven alkalis. Maintaining a slightly alkaline pH condition overall is crucial for having good health. If your body's pH is not balanced, you cannot effectively assimilate vitamins, minerals, supplements, and food. If your pH is too acid, then you are also low in oxygen.

Acid beverages such as soft drinks rob our bodies of oxygen, while alkaline drinks such as alkaline water enrich the body with oxygen and much-needed minerals. Alkaline water also neutralizes free radicals. Alkaline water is the healthiest water.

Many struggle endlessly to remain alkaline because they do not address the direct use of oxygen to release acid toxins and remove the lactic acid in abundance because of low oxygen conditions.

My focus has been on alkalizing the body with sodium bicarbonate, which offers quick control over body pH. Sodium bicarbonate turns to carbon dioxide in the stomach, which secretes hydrochloric acid in response, driving bicarbonates into the bloodstream. One way to increase oxygen delivery to the cells is by increasing bicarbonate and CO2 concentrations, dilating the blood vessels, and ensuring more oxygen.

Metabolic reactions occurring with insufficient oxygen led to acidosis. Thus, hypoxia or poor oxygenation of the tissues is associated with high mortality and can lead to diminished consciousness, cardiac arrhythmias, and subsequent cardiac arrest within minutes. Supplementary oxygen is indicated whenever tissue oxygenation is impaired, such as occurs in COPD) Chronic Obstructive Pulmonary Disease).[174]

Oxygen Controls Alkalinity.

In *The Metabolism of Tumors,* Warburg demonstrated that all forms of cancer are characterized by two necessary conditions: acidosis and hypoxia (lack of oxygen). Lack of oxygen and acidosis are two sides of the same coin: where you have one, you have the other.

According to Keiichi Morishita in his book, *Hidden Truth of Cancer,* if blood starts to become acidic, the body deposits the excess acidic substances into cells to maintain a slightly alkaline condition. This deposit causes those cells to become more acidic and toxic and decreases their oxygen levels.

Over time, he theorizes, these cells increase in acidity, and some die. These dead cells themselves turn into acids. However, some of these acidified cells may adapt to that environment. In other words, instead of dying, as normal cells do in an acid environment, some cells survive by becoming abnormal.

These abnormal cells are called malignant cells. Malignant cells do not correspond with brain function nor with our own DNA memory code. Therefore, malignant cells grow indefinitely and without order. This is cancer.

Alkaline water (including the water in cells) can hold a lot of oxygen. Acidic water (or cells) can carry very little oxygen. So, the more acidic your cells are, the less oxygenated they will be.

If the blood is already too acidic, the body must take the toxins out of the blood and deposit them into cells to keep the blood at the proper pH. Besides, cells cannot release toxins into the blood to detoxify themselves when they are too acidic.

An overload of toxins clogs up the cells, low-quality cell walls that do not allow nutrients into the cells, the lack of nutrients needed for respiration, poor circulation, and low oxygenation levels produce conditions where cells produce excess lactic acid they ferment energy. Lactic acid is toxic and tends to prevent the transport of O2 into neighboring normal cells.[175]

Conclusion.

The human body is alkaline by design but acidic by function. Every living cell in the body creates metabolized waste, which is acidic. The nutrients from our food are delivered to each cell; the cells burn with oxygen to provide energy for us to live. The burned nutrients become metabolized waste, but waste can be recycled and used to balance and increase oxygen levels in carbon dioxide.

All waste products are acid; the body discharges the waste through urine, bile, and sweat. Our body cannot get rid of 100% of the waste it produces all the time, which leads to an overload of toxicity. Without proper elimination, acid waste products become solid wastes, such as micro toxins, toxins, fungus, bacteria, and mucus. These accumulate and build up in our blood, organs, and tissue. This accumulation of waste products accelerates the depletion of minerals and other nutrients, causes disease, and accelerates the aging process. All this drives down healthy oxygen levels into a pit of hypoxic tissues that eventually become cancerous.

The metabolism of cancer cells has a very narrow pH tolerance for cellular proliferation (mitosis) between 6.5 and 7.5. As such, if you can interfere with cancer cell metabolism by either lowering or raising the internal cancer cell pH, you can theoretically stop cancer progression.[176] In my book *Sodium Bicarbonate,* I explore why baking soda is one of the most helpful medicines for cancer. The book's original subtitle was *Rich Man's Poor Man's Cancer Treatment,* and it is about the least expensive heavy-hitting instant-acting medicine out there.

Low oxygen conditions (acidosis) lead to Cardiovascular damage. Weight gain, obesity, and diabetes. Bladder condition. Kidney stones. Immune deficiency. Acceleration of free radical damage. Hormonal problems. Premature aging. Osteoporosis and joint pain. Aching muscles and lactic acid build-up. Low energy and chronic fatigue. Lower body temperature. A tendency to get infections. Loss of drive, joy, and enthusiasm. Depressive tendencies. Easily stressed. Pale complexion. Headaches Inflammation of the corneas and eyelids. Loose and painful teeth. Inflamed sensitive gums. Mouth and stomach ulcers. Cracks at the corners of the lips. Excess stomach acid. Gastritis. Nails are thin and split easily. Hair looks dull, has split ends, and falls out. The skin is easily irritated. Cramp and spasms.

Acid Bodies; Acid Cancer.

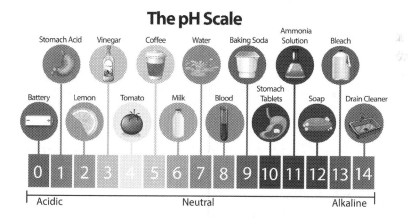

The pH Scale

Cancer involves an interaction between rogue cells and surrounding tissue. This is the clear message of Dr. Mina Bissell. The health or sickness of surrounding cells and the surrounding extracellular matrix interact to shape cancer cell behaviors such as polarity, migration, and proliferation.

Cancer cells routinely form in most people's bodies in areas of low voltage, low oxygen, and acidic pH. Bottom line: the more acid conditions prevail, the more aggressive the cancer. Hypoxia and extracellular acidity are deeply associated with the cellular microenvironment and the spread of cancer.

The direct effect of oxygen on pH on the water is trivial, less than temperature variations; however, the indirect effects are immense.

Without sufficient levels of oxygen, the cells cannot detoxify themselves, so acids build up. Without adequate oxygen, the cells turn to fermentation as an alternate energy source for survival, which builds up acidity by creating lactic acid. Without enough oxygen, cells turn cancerous, or they die.

The term "pH" is used within chemical formulas to indicate the presence of hydrogen. The "H" means hydrogen in this formula, just as the "H" in "H2O" suggests the presence of hydrogen in water -- two atoms of hydrogen and one atom of oxygen.

A heavily acid solution can have one hundred trillion times as many hydrogen ions as an alkaline solution. To deal with these vast numbers, scientists use a logarithmic scale known as the pH scale. Each number on the scale indicates a change of 10 times the concentration of hydrogen ions in the solution. Hydrogen in its molecular form and H- ions are healthy, but all those H+ ions are a bitch.

In general, an increase in hydroxyl ion concentration increases oxygen consumption. Higher pH means a higher concentration of oxygen, while lower pH means lower concentrations of oxygen.

Increasing pH from 4.0 pH to 5.0 pH increases oxygen molecules in a liquid by ten-fold. Each whole number increases by tenfold again, so 4.0 pH to a 6.0 pH increases oxygen by 100 times, and raising pH from 4.0 pH to 7.0 increases oxygen levels 1,000 times.

PH is the measurement of the oxygen to hydrogen ratio in a liquid, ranging from 14 (a lot of oxygen – or Alkaline) to 1 (a lot of hydrogen – or Acidic). Bill Farr writes, "If you think of your cells as mini-vans, carrying ten passengers, as long as seven of them are oxygen, and only three are hydrogen, all will be healthy and happy. That is until the day you decide you want a coke, some fried or processed foods instead of vegetables, and continue eating this way. Soon, all hydrogen ion passengers push the oxygen molecules out of the van, and your pH level starts to drop. Now, your body becomes desperate to pick up oxygen, but only hydrogen is offered, sliding the pH scale down to acidic. Then these hydrogen passengers begin to invite their friends to the party. Those friends are cancer, insomnia, bacterial infections, food allergies, and the list goes on."

An acidic pH can occur from an acid-forming diet that does not provide all the necessary minerals, emotional stress, toxic overload, immune over-reactions, or any process that deprives the cells of oxygen and other nutrients. Normal body cells form lactic acid when their oxygen is cut off or their respiration is reduced. Scientists have found that during a forced workout, the lactic acid of the blood increased. In this case, oxygen diffusion into the muscle cells is insufficient to cover the muscle's oxygen requirement.

Blood's pH comes into play because CO2 concentration helps determine pH. Carbon dioxide reacts with water to form carbonic acid; this reaction is catalyzed or accelerated by an enzyme found in our blood called carbonic anhydrase. Carbonic anhydrase loses a hydrogen ion to become bicarbonate, and this, in turn, loses a hydrogen ion to become the carbonate ion.

This chain of reactions is two-way. Carbonate can pick up a hydrogen ion to become bicarbonate and vice versa. pH is a measure of the hydrogen ion concentration in a solution, so this chain of reactions plays a critical role in maintaining or buffering our blood pH—the more positive hydrogen ions, the less oxygen.

The Henderson-Hasselbalch equation shows that pH is governed by the ratio of base (HCO3−) concentration to acid (H2CO3) concentration as hydrogen ions are added to the bicarbonate buffer.

$$H+ \text{ plus } HCO3 = H2CO3$$

Thus, the bicarbonate (base) is consumed (concentration decreases), and carbonic acid is produced (concentration increases). Continue to add hydrogen ions; all bicarbonate eventually is consumed (converted to carbonic acid), and there would be no buffering effect – pH then falls sharply if more acid is added.

However, if carbonic acid could be continuously removed from the system and bicarbonate regenerated continually, then the buffering capacity and, therefore, pH could be maintained despite the continued addition of hydrogen ions.

Crucial pH.

A slight acidity depresses oxidation, and increasing alkalinity significantly increases the oxidation rate even to a marked degree. Warburg, McClendon, and Mitchell showed that an increase in hydroxyl ion concentration increased oxygen consumption. Consumption of oxygen in the tissues increases with increasing pH.[177]

The pH of our blood is very tightly buffered thanks to the bicarbonate it contains and to hemoglobin. That is why the pH of our blood stays within a narrow range. Since our cells release carbon dioxide as they break down sugars, the carbon dioxide and carbonic acid concentration are higher in blood flowing through your tissues than blood in your lungs, where is relatively poor carbon dioxide. Consequently, the pH of blood in the lungs stays relatively constant at around 7.6, while the pH of blood in the tissues is closer to 7.2. This slight difference in pH has significant ramifications.

Signs of Low Oxygen Acidic Conditions.

The essential factor in creating proper pH is increasing oxygen. Suppose your body's pH is not correct because of low oxygen conditions; in that case, you cannot effectively assimilate vitamins, minerals, supplements, and food. That is why, even with a good diet, it is often hard to pull out of diseased conditions. Our body's pH affects everything, including oxygen transport and delivery to the cells.

Ed McCabe, famously known as Mr. Oxygen, wrote, "Alkalinity and Oxygen go hand in hand. Being healthy, alive, and burning food for energy includes the constant process of billions of tiny cells making chemical reactions that continually generate waste acids in our bodies. The constant cleaning of the waste acids as fast as we make them is paramount to good health. Our bodies must maintain the proper balance between being alkaline or acid pH for us to be alive."

Beware.

Some people try to debunk the entire acid-alkaline story of human health and disease, saying, "Because the cancer cell is producing lactic acid, does not mean that the whole body becomes acidic. The body has a very tight set of checks and balances, which keeps your blood's (i.e., the body's) pH between 7.35 – 7.45 or near neutral. It is also impossible to significantly alter one's blood pH with the food you eat. Therefore, checking your saliva for acidity or alkalinity is not an accurate measure of your body's actual pH balance (in the blood)." Still, it does give one an idea of the state of balance of the rest of the body.

Conclusion.

No wastes or toxins can leave the body without first being combined with oxygen, so you become more toxic if you are not getting enough O2. The more toxins, the less oxygen, the less oxygen, the more acidity, and the more toxicity builds up. A body with too many accumulated waste toxins and acids is unhealthy and often painful. As readily seen – and felt – an example of this is when we become sore after exercise. This soreness is entirely due to the over-accumulation of lactic acid in our muscles. It is just waiting to be removed. In a few minutes, hours, or days, the body marshals its resources and buffers the lactic acid – and the soreness goes away.

Cancer causes an over-accumulation of lactic acid around the cancer site that causes other nearby cells to become aberrated. Some remove this cascade of advancing acid-fueled cancers by increasing alkalinity with sodium bicarbonate.

Oxygen can be Dangerous but is Vital.

While it is generally safe, Oxygen Therapy carries the risk of complications that can be life-threatening and/or result in permanent or long-term disability in rare instances. According to new Canadian research, oxygen is given to millions of patients worldwide every day, but too much of it can be harmful. Published in April 2018 in The Lancet, presented evidence shows supplemental oxygen, when given liberally to acutely ill patients, increases the risk of death in people with sepsis, stroke, cardiac arrest, and those with trauma or requiring emergency surgery.

Most people believe that oxygen therapy is harmless when it is not. These medical scientists concluded that an additional death occurs in hospital for every 71 patients treated with excessive oxygen, suggesting the need to move away from being too liberal with oxygen," said the study's lead author, Dr. Derek Chu, a clinical fellow at McMaster University.

The study found that supplemental use of oxygen:
Did not decrease any infection.
Did not improve a patient's length of hospital stay.
And for those with a stroke or brain injury, it did not improve a patient's level of disability.

In 1947, the British military discovered that oxygen could be toxic (Oxygen Toxicity) due to underwater research. The same study found that nitrogen (78% of air) caused Nitrogen Narcosis. Oxygen toxicity is a condition resulting from the harmful effects of breathing molecular oxygen at increased partial pressures. It is also known as oxygen toxicity syndrome, oxygen intoxication, and oxygen poisoning. Severe cases can result in cell damage and death, with effects most often seen in the central nervous system, lungs, and eyes.

The result of breathing increased partial oxygen pressures is hyperoxia, an excess of oxygen in body tissues. The body is affected in diverse ways depending on the type of exposure. Central nervous system toxicity is caused by short exposure to high partial oxygen pressures greater than atmospheric pressure. Pulmonary and ocular toxicity results from prolonged exposure to increased oxygen levels at standard pressure. Symptoms may include disorientation, breathing problems, and vision changes such as myopia. Prolonged exposure to above-normal oxygen partial pressures, or shorter exposures to very high partial pressures, can cause oxidative damage to cell membranes, the collapse of the alveoli in the lungs, retinal detachment, and seizures.

"There's no clear consensus in North America or around the world on how to use oxygen," says Dr. Chu, meaning there is quite a bit of medical ignorance surrounding the most prescribed drug used in hospitals. Obviously, as demonstrated by all manufacturers and suppliers of oxygen cylinders, CO_2 is always added because CO_2 makes oxygen safer, as mentioned many times in this book. The subtitle reveals medicine's future and the safe administration of oxygen—Combining Oxygen with Hydrogen and CO_2.

Hydrogen Medicine is relatively new to the world of medicine. It will take doctors out of the dark ages of pharmaceutical medicine and give birth to new understandings of how medical gases need to be used to increase safety and effectiveness in serious disease surgery.

Cardiologist John William McEvoy believes the results will lead to a reconsideration of guideline recommendations around supplemental oxygen, which in the end should always include mixing hydrogen with oxygen. McEvoy, an assistant professor at John Hopkins University School of Medicine, said the new research would change how he administers oxygen in his coronary care unit. "This is definitely a practice-changing study. I read very many papers and meta-analyses, and this is one of the few that I think truly should change how we think about oxygen."

Oxygen is often administered because it is widely believed to be safe and not considered a harmful substance. The most common reason to give extra oxygen is for short-term breath patients with lung conditions or oxygen-deficient. "It's just common practice — a traditional practice — that we give them some excess oxygen because they're in the ICU, they're intubated, and we don't think it's harmful," McEvoy said. It is common because it should be; most seriously ill people are oxygen-deficient, which goes double for cancer patients. What needs to be decided is the best way of administering oxygen, which this book addresses as one of its main points.

Oxygen is an Extremely Important Subject.

For several years, atmospheric oxygen (O_2) has been dropping faster than the amount that goes into the increase of CO_2 from burning fossil fuels, some 2 to 4-times as much, accelerating since 2002-2003. Simultaneously, the ocean oxygen levels worldwide are falling.

It is becoming clear that getting rid of CO_2 is not enough; oxygen has its own dynamic; the rapid decline in atmospheric O_2 must also be addressed. Although there is much more O_2 than CO_2 in the atmosphere - 20.95 percent (209,460 ppm) of O_2 compared with around just over 400 ppm of CO_2– humans, all mammals, birds, frogs, butterfly, bees, and other air-breathing life-forms depend on this elevated level of oxygen for their well-being. In humans, failure of oxygen energy metabolism is the most critical risk factor for chronic diseases, including cancer and death. 'Oxygen deficiency' is currently set at 19.5 percent in enclosed spaces for health and safety. Below that, fainting and death may result.

It is not difficult to diagnose oxygen deficiency. All one needs to do is slip on an inexpensive oximeter on one's finger to see the blood's oxygen levels. One can also count one's breathing rate, for the faster one breathes, the more CO_2 a person blows off, the lower the oxygen levels fall. When chronically ill or with cancer, one will never recover without recovering from low oxygen levels. An important note for both patients and clinicians is that oxygen levels can fall to dangerously low levels in just one part of the body while the rest remains normal. A cancer patient can show high oxygen levels in the blood (which is what oximeters test), yet the prostate gland or rectum tissues or one's toes can below.

Oxygen therapy will become even more important as the years go by; it is already a critical issue for many. This is as true for newborns and infants as it is for older people suffering from chronic illness. So, it behooves us to understand how oxygen can be administered in manners that will not increase death risk.

Oxygen can be safely administered differently from what hospitals are presently doing. The above research points to changes in oxygen administration, but doctors and medical institutions do not know which changes to make. In my book *Anti-Inflammatory Oxygen Therapy*, I champion exercise with oxygen therapy (EWOT), which is safe because of the massive increase in CO_2. Oxygen is deadly without CO_2, and yet in the above study of hospital use of oxygen, CO_2 is not considered a crucial factor.

The level of carbon dioxide (CO_2) in our body is what controls our breathing. When carbon dioxide reaches a certain level, the breathing center in your brain stem sends a signal to the breathing muscles, triggering an inhalation. We exhale carbon dioxide, and a new breathing cycle starts. Carbon dioxide is produced continuously in your body when you breathe; you exhale the built-up CO_2. The more active we are, the more CO2 is produced. That is why we breathe more when we are out jogging than when we sit on the sofa relaxing.

EWOT is safe but not practical for ICU, surgery, or emergency rooms. Singlet oxygen administration and ozone oxygen therapy are introduced at the end of the book, with singlet oxygen safe because more oxygen is not consumed.

Hydrogen makes oxygen safer for several reasons. First, we need less oxygen for healing when hydrogen is present. That is illustrated by the fact that divers at 2000 feet below sea level breathe up to 96 percent hydrogen and only 4 percent oxygen. Hydrogen puts out the oxidative fires inherent in the body's oxygen use, turning the nastiest free radicals into water.

One of my favorite ways of increasing oxygen to the cells is to slow the breathing down, which increases CO2/bicarbonates in the blood. The other is using sodium bicarbonate, which increases oxygen almost instantly.

I have written about hyperbaric oxygen therapy in my *Anti-Inflammatory Oxygen Therapy* book, and it is safe when used correctly. Yet we read: "Sudden death during hyperbaric oxygen therapy: rare, but it may occur."[178] The former medical director of the clinic, where a hyperbaric chamber explosion killed a four-year-old boy and his grandmother in 2009, was sentenced to two years of house arrest. Things happen, though, even when you step out the door.

One will find all kinds of nasty things said about ozone therapy because mainstream medicine is threatened by it. I trust the word and work of Dr. David Brownstein, who says, "Ozone therapy, when done correctly, is one of the most effective and safest therapies I have seen. The big danger of ozone therapy is inhaling it--as long as that doesn't happen, it is a safe therapy. Ozone can be injected, insufflated in the rectum, vagina, or blown into the ears safely. When used correctly, the only real side effect I see from ozone therapy is a Herxheimer reaction as bugs get killed from ozone."

Hydrogen, especially when mixed with oxygen, is potentially explosive, so holding a match to one's nose is not a clever idea when doing hydrogen inhalation. Eventually, there will be an accident, for even static electricity can cause an explosion. That is why I evaluate all hydrogen inhalers and recommend those that seem safest. Suppose sellers of hydrogen inhalers do not tell the buyer of the dangers. In that case, there is potential criminal liability resulting in jail time, especially if the seller knew of the threat and did not divulge it.

Everything said above needs to be understood in the proper context of medicine and medical practice, where even taking aspirin can kill you. Over 20,000 Americans die from all the different aspirin types, with Tylenol sending more people to the emergency room than just about any other substance. Oxygen is only dangerous when used without intelligence, and though in certain conditions, hydrogen mixed with oxygen can be explosive, it rarely is when used in medicine.

When reviewing the differences between inhalation of hydrogen and drinking hydrogen water, we must see that inhalation does not help the gut recovery as much as hydrogen water does (most people nowadays have major gut issues). When hydrogen water gets in the gut, it selectively stimulates anaerobic microflora, and gastric Ghrelin is stimulated. This is highly beneficial even with neurological problems, for so much depends on the health of the gut. Thus, it is almost always recommended that hydrogen inhalation be accompanied by hydrogen water administration, especially over the long term.

Hydrogen inhalation is much more flexible, with unlimited possibilities for serious life-threatening medical situations like cancer, sepsis, emergencies, ambulances, surgery, and in the ICU. When a person is dying, one can simply hook them up to a hydrogen (or hydrogen-oxygen) machine and let it run continuously until the patient responds positively. One cannot do this with water, especially when a patient cannot drink water even though they should.

I recently attended a small girl in Brazil mistreated by what hardly would pass for a hospital in the first world. She was suffering from a severe gut problem and blinding pain. They just gave two shots for pain (without regard at all to the girl's low body weight) and injectable antibiotics. Neither helped. Being a neighbor and friend of our 13-year-old daughter, we invited her to stay with us and administered hydrogen and oxygen gas continually. Instead of hydrogen water, we had her sip water with sodium bicarbonate, potassium bicarbonate, and magnesium carbonate (product is called pH Adjust), alternating with sipping water with iodine to chase down a not too good tasting but excellent and powerful selenium. Within two hours, the patient looked and felt better. Within a day, she was eating a little again, and the next day was ready to go home.

Hydrogen-led protocols provide us with an entirely new concept in intensive care treatment delivered at home or in ICU departments. As the above case exemplifies, hydrogen leads the therapy but needs to be balanced and supplemented with other essential medicines. Hydrogen should not be seen or used alone but always in the context of a complete protocol.

Ozone Therapy and the Oxygen Wars.

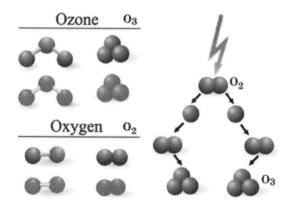

"The federal and state medical authorities are currently seizing and destroying ozone machines. The legal conventions here assume that a physician can't use a treatment that the other doctors around him aren't using. They came in and arrested him, trying to strip him of his license. Many cases are pending, sincere ' doctors trying to help their patients by using harmless natural remedies. They're being attacked, losing their insurance privileges, and they won't let them use ozone. I interview cases where they use SWAT teams to break into a home and pushed everyone, mothers and grandmothers, against the wall at gunpoint and take all the patient records, the ozone machine, and the computers, including backups. The doctors never get them back. Certain newspapers report this but leave out the indignities. In their stories, they ignore stacks of medical evidence sent to them - including the personal testimony of ex-patients who show up in their editorial offices to correct the biased reporting. They write, "Experts say ozone is worthless." "Experts" with absolutely no training or experience with medical ozone. But the public does not know that. We are witnessing the oxygen wars," writes Ed McCabe.

Ozone removes viruses and bacteria from the blood. It has successfully been used on AIDS, herpes, hepatitis, mononucleosis, cirrhosis of the liver, gangrene, cardiovascular disease, arteriosclerosis, high cholesterol, cancerous tumors, lymphomas, leukemias, highly effective on rheumatoid and other arthritis, improves mental sclerosis, ameliorates Alzheimer disease, senility, and Parkinson's, effective on proctitis, colitis, prostate, candidiasis, trichomoniasis, and cystitis. Externally, ozone effectively treats acne, burns, leg ulcers, open sores and wounds, eczema, and fungus. That is the kind of world we live in, where we have SWAT teams trying to eliminate good things from life. Ozone is a form of oxygen, an exciting form with interesting broad uses.

During WWI, physicians applied ozone topically to infected wounds because of its antibacterial properties. Later they discovered it had broader therapeutic applications, thanks to its anti-inflammatory properties. In the late 1980s, German physicians began successfully treating HIV patients with ozone. Although some practitioners administer ozone, it remains an uncommon treatment within the U.S. even today. Ozone is a well-respected therapy in many parts of the world. In Germany it is the standard of care practice and is used by 70-80% of practicing physicians.

The lack of health that exists in every single cancer condition, without exception (no matter the location or stage of cancer), has to do with impaired, oxygen-less respiration of the body's cells. Ozone, like its progenitor, oxygen, is a gas. Oxygen (known chemically as O_2) likes to travel in pairs.

When the third atom of oxygen binds, O_2 becomes O_3. Because O_3 is inherently unstable, it always wants to give away that extra atom. At the same time, any cell that encounters O_3 will take this third atom. When this happens, oxygen's traditional properties become more powerful and more energized.

Ozone's most unique property is that it is a potent oxidant. It will break down any chemical into that chemical's basic parts. Ozone has been used medically to disinfect and treat disease since its discovery. In 1896, Nikola Tesla patented the first ozone generator in the United States. Ozone has been used as a safe and adequate water purifier for more than a century. Ozone deactivates pathogenic microbes in the human body in much the same way it does in water, unsurprising since our bodies are made up of 70% water.

Dr. Frank Shallenberger wrote, "You may have heard me report to you on several occasions about how amazing ozone therapy is at helping your body to heal and stay well. You probably know that I've been teaching doctors from all over the world for more than 25 years to use ozone therapy on their patients. But wouldn't it be great if you could use many of the amazing healing properties of ozone therapy right in your own home? Here is the good news: You can and do not have to be a doctor. You do not even have to have any medical experience at all. If you can make a cup of coffee, you can learn to treat yourself and your family with ozone."

Inflammation, Oxygen, CO2 & Breathing.

The complicated world of oxygen, carbon dioxide, and tissue and tumor pH is important because our body simply cannot fight disease if its pH is not balanced correctly. Consequently, the oxygen-carrying capacity of our cells becomes compromised. It is straightforward—higher pH conditions lead to higher O2 levels resulting in oxygen being delivered to where it is needed.

What follows are lessons on CO2 physiology and breathing. CO2 physiology is super important and more complicated than any other health factor, so beware and prepare for a deep journey into places where few have gone before. After, we will go back to more on oxygen and a brilliant and powerful way of combining CO2 therapy with massive oxygen increases. Exercise with Oxygen Therapy (EWOT) offers the best of both oxygen and CO2.

The key to oxygen is not more oxygen but more carbon dioxide, a nutritious gas, not a poison. Doctors at the Department of Anesthesia and Medical-Surgical Intensive Care Unit, Toronto General Hospital in Ontario, Canada, say that "Accumulating clinical and basic scientific evidence points to an active role for carbon dioxide in organ injury, in which raised concentrations of carbon dioxide are protective and low concentrations are injurious."[179]

Carbon dioxide executes uncountable functions in the human organism. Among them are repair of alveoli in lungs, the stability of the nerve cells, regulation of pulse, normal immunity, blood pressure maintenance, dilation of bronchi and bronchioles, regulation of blood pH, sleep control, relaxation of muscle cells, the release of O2 in capillaries (the Bohr effect), and other essential functions.

Hemoglobin helps to transport hydrogen ions and carbon dioxide in addition to transporting oxygen. However, hemoglobin accounts for only 14% of the total transport of these species; both hydrogen ions and carbon dioxide are also transported in the blood as bicarbonate (HCO3-).

Sodium bicarbonate (baking soda) is a stunning medicine because it puts doctors' and patients' fingers on the CO2 pulse of the body. Bicarbonate intake raises the CO2 levels in the blood. CO2 is a key regulator of inflammatory reactions due to the control of cell oxygen supply. Bicarbonate also regulates inflammatory responses due to rapid changes in tissue and fluid pH. Bicarbonate and CO2 are almost identical twins, two sides of the same coin. Just add acid to bicarbonate, and you instantly have CO2. With the help of anhydrase, they turn into each other in the blood.

> *In all serious disease states, we find a concomitant low-oxygen state. Low oxygen in the body tissues is a sure indicator for disease. Hypoxia, or lack of oxygen in the tissues is the fundamental cause of all degenerative diseases.*

> Dr. Stephen Levine
> Molecular Biologist.

The effect of bicarbonate is instant and can be intense, as an athlete might tell you when taking bicarbonate before an event. To survive, the body must maintain proper acid/alkaline (pH) balance. The optimum (and required) pH of the blood is alkaline, between 7.35 and 7.45. Only in this range is the blood richly supplied with oxygen.

Realize that if your blood pH varies just a little bit, it can kill you. Most terminal cancer patients are more acidic than healthy people, meaning some of their tissues have a low pH and low oxygen delivery levels to the cells. This is true even though the pH of the blood is controlled to preserve life. However, even the slightest change of the blood pH is meaningful in terms of oxygen.

Natural Allopathic Medicine Protocol.

Natural Allopathic Medicine targets a sweet spot—that area of pH and O2 that healthy cells love and cancer cells don't like at all. Oxygen levels are sensitive to a myriad of influences. Toxicity, emotional stress, physical trauma, infections, reduction of atmospheric oxygen, nutritional status, lack of exercise, and improper breathing will affect our bodies' oxygen levels. Anything that threatens the oxygen-carrying capacity of the human body will promote cancer growth. Likewise, any therapy that improves oxygen function can be expected to enhance the body's defenses against cancer.

Low Carbon Dioxide Leads to Cancer.

Under clinical conditions, low oxygen and low carbon dioxide occur together. Therapeutic increase of carbon dioxide, by inhalation of this gas diluted in air, is often an effective means of improving the blood and tissues' oxygenation.[180]

Carbon dioxide is one of the most important gases for life. It is healthy and indispensable to our biological existence. CO_2, the waste product of cell metabolism, is not waste at all. Plants thrive on it, and our lives depend on it.

Dr. Buteyko said, "CO_2 is the main source of nutrition for any living matter on Earth. Plants obtain CO_2 from the air and provide the main source of nourishment for animals, while both plants and animals are nourishment for us. The great resource of CO_2 in the air was formed in pre-historical times when the amount was about 10%."

Everything is toxic when it is too much, and that is true even for water and oxygen. In the air that we breathe, current levels of CO_2 are not even close to dangerous. Medically speaking, the real problem is when there is not enough CO_2—when we do not exercise enough or when we breathe too fast, we tend to drive down CO_2 levels in the blood.

Improper fast breathing (which is the norm today) causes oxygen deficiency because we are ventilating too much CO2, which contracts the blood vessels and changes the oxygen disassociation curve in a way that slowly suffocates our cells. Hypocapnia (lowered CO2) reduces oxygenation of all vital organs and tissues due to vasoconstriction and the suppressed Bohr effect.

CO2 and bicarbonate, carbon dioxide's twin sister, are the vital players in the pH balance in cells, blood, and other bodily fluids meaning CO2 holds the keys to oxygen delivery. If the blood's carbon dioxide level is lower than normal, this leads to difficulties in releasing oxygen from hemoglobin.

Poor oxygenation or hypoxia is a favorable environment for cancer development, whereas adequate oxygenation favors healthy tissue growth. Increasing Co2 levels through sodium bicarbonate is helpful in cancer treatment because bicarbonate drives CO2 levels in the blood, improving oxygenation to the cells.

Most doctors ignore CO2, even though lowered carbon dioxide levels in the blood lead to reduced oxygenation of the cells, which leads to toxic accumulation and increased acidity. This condition has cancer written all over it. In the presence of a large amount of carbon dioxide, the hemoglobin molecule changes its shape slightly, favoring oxygen release.

If a carbon dioxide deficiency continues for a long time, it can be responsible for diseases and accelerated aging. As physiological studies have found, hypocapnia constricts blood vessels and decreases all vital organs' perfusion. In emergencies, when these conditions are extreme, the body's organs and tissues are not receiving an adequate flow of blood.

Hypoxia Suppresses the Immune System.

"Hypoxia and immunity are highly interdependent. Hypoxia affects molecular and cellular inflammatory processes. Hypoxia activates distinct hypoxia-signaling pathways, including a group of transcription factors known as hypoxia-inducible factors and adenosine signaling. In vitro and animal studies have shown that these pathways participate in the modulation of inflammatory responses. Inflammatory conditions are frequently characterized by tissue hypoxia due to enhanced metabolic demand as well as decreased metabolic substrates resulting from edema, microthrombi, and atelectasis, in turn causing "inflammatory hypoxia."[181]

Hypoxia Supports Tumor Growth.

Hypoxia interferes with effective radiation and chemotherapy. Hypoxia incapacitates several diverse types of immune effector cells, enhances immunosuppressive cells' activity, and provides new avenues that help "blind" immune cells to the presence of tumor cells.[182] Hypoxia is the enemy of the anti-tumor immune response. Oxygenation, on the other hand, would reduce tumors' escape from immune surveillance and response.

Several doctors report, "Rapidly growing tumors with poorly formed vasculature have low oxygen levels, and limited oxygen availability results in a hypoxic microenvironment. Patients with elevated levels of tumor hypoxia have a significantly worse prognosis than patients with low levels. Thus, targeting tumor hypoxia in the treatment of prostate cancer has the potential to improve patient response to treatment and overall survival."

Hypocapnia (CO_2 deficiency) in the lungs and, in most cases, arterial blood is a normal finding in chronic diseases due to the prevalence of chronic hyperventilation among the sick. Understanding the disease's pathogenesis, in which hypocapnia is a constitutive element, is necessary to understand cancer. Hypocapnia is a universal constant behind the condition.

The Warburg effect (WE), or aerobic glycolysis, triggered by hypoxia, is commonly recognized as a hallmark of cancer and has been extensively studied for potential anti-cancer therapeutics development.[183]

Oxygen, Inflammation & Hypoxia-Inducible Factor

In Germany, scientists have shown that the microenvironment of inflamed and injured tissues is typically characterized by low oxygen and glucose levels and elevated levels of inflammatory cytokines, reactive oxygen, and nitrogen species and metabolites. Medical research suggests a strong link between cell hypoxia (oxygen deficiency in cells) and chronic inflammatory processes.

Inflammation is the most common cause of tissue hypoxia and decreased circulation. Both inflamed tissues and the areas surrounding malignant tumors are characterized by hypoxia and low concentrations of glucose. Inflammation can lead to sepsis, circulatory collapse, and ultimately to multi-system organ failure.

Tissue hypoxia is manifested in increased levels of hypoxia-inducible factor (HIF-1) (this factor and cell hypoxia are vital factors in the progress of cancer). Elevated HIF-1 triggers a cascade of events involving pro-inflammatory transcription factors such as nuclear factor kappa B (or NF-kappaB) and activator protein AP-1.

Researchers have found that low levels of magnesium suppress reactive oxygen species (ROS) induced HIF-1. When oxygen levels fall, things get dangerous on a cell level because gene expression changes at low levels. HIF-1a regulates the expression of at least 30 genes when oxygen levels are low. Magnesium deficiency depresses HIF-1a activity.

Often an excessive inflammatory immune response (sepsis) contributes to a patient's death. In intensive care units, sepsis is the most common cause of death. Patients with a severely compromised immune system face attacks from Candida fungal infections, which become life-threatening because of the high risk of sepsis.

Cancer & HIF-1.

"Radiation and chemotherapy do kill most solid tumor cells, but in the cells that survive, the therapies drive an increase in HIF-1, which cells use to get the oxygen they need by increasing blood vessel growth into the tumor. Solid tumors generally have low supplies of oxygen, and HIF-1 helps them get the oxygen they need," explains Dr. Mark W. Dewhirst, professor of radiation oncology at Duke University Medical Center.

Dr. Holger K. Eltzschig, a professor of anesthesiology, medicine, cell biology, and immunology at the University of Colorado Medicine School, says, "Understanding how hypoxia is linked to inflammation may help save lives. By focusing on the molecular pathways the body uses to battle hypoxia, we may be able to help patients who undergo organ transplants, who suffer from infections or have cancer."

Researchers found that an increase of 1.2 metabolic units (oxygen consumption) was related to a decreased risk of cancer death, especially in lung and gastrointestinal cancers.[184]

For cancer to "establish" a foothold in the body, it must be deprived of oxygen and become acidic. If these two conditions can be reversed, cancer not only can be slowed down but can also be reversed.

Drs. D. F. Treacher and R. M. Leach write, "Prevention, early identification, and correction of tissue hypoxia are essential skills. If oxygen supply fails, even for a few minutes, tissue hypoxemia may develop, resulting in anaerobic metabolism and production of lactate."[185]

The Bohr Effect.

In 1904, Danish scientist Christian Bohr noticed that hemoglobin binds oxygen more tightly at high pH than at low pH. The Bohr effect explains oxygen release in capillaries, why red blood cells unload oxygen in tissues. Bohr stated that hemoglobin would bind to oxygen with less affinity at lower pH (more acidic environment, e.g., in tissues). The Bohr effect has to do with hemoglobin's ability to pick up or donate hydrogen ions. As pH rises, hemoglobin loses hydrogen ions from specific amino acids at key sites in its structure. This causes a subtle change in its structure that enhances its ability to bind oxygen.

When pH falls in the blood, when it becomes slightly more acidic, the reverse happens: hemoglobin picks up hydrogen ions, and its affinity for oxygen decreases. Your blood's pH is very tightly buffered thanks to the bicarbonate it contains and hemoglobin, which can pick up or lose hydrogen ions to counteract pH changes. Hemoglobin affinity for oxygen is how readily hemoglobin acquires and releases oxygen molecules into the fluid surrounding it.

Hemoglobin will drop off more oxygen as carbon dioxide concentration increases dramatically, like when we exercise, when tissue respiration is happening rapidly, and oxygen is in greater need. The opposite is true under low CO_2 levels; hemoglobin will drop off less oxygen. Increasing CO_2 concentration drives a decrease in pH, which helps force hemoglobin to dump the oxygen it is carried from the lungs, so your cells can use it to break down sugars for energy. It decreases CO_2 through faster than normal breathing when at rest that has the opposite effect of putting the brakes on oxygen delivery. The pH-mediated change in affinity for oxygen helps hemoglobin act like a shuttle that picks up oxygen in the lungs and deposits it in the tissues where it will be needed.

The dissociation of oxygen is also helped by magnesium because it provides an oxygen adsorption isotherm, which is hyperbolic. It also ensures that the oxygen dissociation curves are sigmoidal, maximizing oxygen saturation with the gaseous oxygen pressure (Murray et al. pp. 65-67).

Oxygen dissociation with increased delivery to the tissues is increased by magnesium by elevating 2,3-bisphosphoglycerate/DPG (Darley, 1979). Magnesium stabilizes the ability of the phorphyrin ring to fluoresce. The free-radical attack of hemoglobin yields ferryl hemoglobin [HbFe4+] (D'Agnillo and Alayash, 2001), which is inhibited by magnesium (Rock et al., 1995).

Arterial hypocapnia (CO2 deficiency) causes Tissue Hypoxia.

Cell hypoxia is one of the leading causes of free radical generation and oxidative stress leading to inflammation, especially in the capillaries. Capillaries are critical determinants of oxygen and nutrient delivery and utilization, so inflammation there is telling.

It just so happens that normal arterial levels of CO2 have antioxidant properties. A group of Russian microbiologists discovered that "CO2 at a tension close to that observed in the blood (37.0 mm Hg) and high tensions (60- or 146-mm Hg) is a potent inhibitor of the generation of the active oxygen forms (free radicals) by the cells and mitochondria of the human and tissues."[186]

Dr. L.O. Simpson asserts that Fatigue Immune Deficiency Syndrome (CFIDS) results from "insufficient oxygen availability due to impaired capillary blood flow." Tissue oxygenation is severely disturbed during pathological conditions such as cancer, diabetes, coronary heart disease, stroke, etc., associated with a decrease in pO2, i.e., 'hypoxia.'[187] Oxygen delivery is dependent on the metabolic requirements and functional status of each organ. Consequently, organs and tissue are characterized by their own unique 'tissue normoxia' or 'physioxia' status in a physiological condition.

Biologist Dr. Ray Peat tells us that, "Breathing pure oxygen lowers the oxygen content of tissues; breathing rarefied air, or air with carbon dioxide, oxygenates and energizes the tissues. If this seems upside down, it is because medical physiology has been taught upside down.

Respiratory physiology holds the key to all the organs' special functions and many of their basic pathological changes."[188]

Every cell in our body can recognize and respond to changes in the availability of oxygen. The best example of this is when we climb to high altitudes where the air contains less oxygen. The cells recognize the decrease in oxygen via the bloodstream and react, using the 'hypoxic response' to produce an EPO protein (erythropoietin). This protein stimulates the body to produce more red blood cells to absorb as much reduced oxygen levels as possible.[189]

Conclusion.

Carbon dioxide, like air, water, and oxygen, is essential for life, health; it holds the key to resolving asthma, cancer, and many other chronic diseases. Carbon dioxide is an essential constituent of tissue fluids and should be maintained at an optimum level in the blood. The gas is needed to supplement various anesthetic and oxygenation mixtures for special conditions such as cardiopulmonary bypass surgery and the management of renal dialysis. Red blood cells have been reported to shrink and become stiffer under hypoxic conditions[190] and lose their optimal shape when magnesium deficient.

Life Depends on Carbon Dioxide.

Dr. Ray Peat says, "Breathing too much oxygen displaces too much carbon dioxide, provoking an increase in lactic acid; too much lactate displaces both oxygen and carbon dioxide. Lactate itself tends to suppress respiration. Oxygen toxicity and hyperventilation create a systemic deficiency of carbon dioxide. Carbon dioxide deficiency makes breathing more difficult in pure oxygen, impairs the heart's ability to work, and increases blood vessels' resistance, impairing circulation and oxygen delivery to tissues. In conditions that permit greater carbon dioxide retention, circulation is improved, and the heart works more effectively. Carbon dioxide inhibits the production of lactic acid, and lactic acid lowers carbon dioxide's concentration in a variety of ways."[191]

Carbon dioxide is a nutrient and a product of respiration and energy production in the cells. Its lack or deficiency is a starting point for different disturbances in the body. Carbon dioxide has protective functions, including increasing Krebs cycle activity, which is the key to health and the greatest way of avoiding cancer. This is one of the principal reasons bicarbonates are essential medicines for cancer patients, increasing the amount of CO_2 reaching the mitochondria.

CO_2 inhibits toxic damage to proteins. Carbon dioxide is a harmless, colorless, non-toxic natural gas that is the key link in life's carbon cycle. Increasing carbon dioxide inhibits lactic acid formation, thus helps control systemic acidification, which decreases oxygen utilization. CO_2 has led to better coordination of oxidation and phosphorylation and increased the phosphorylation velocity in liver mitochondria.

People who live at very high altitudes live significantly longer; they have a lower incidence of cancer and heart disease, and other degenerative conditions than people who live near sea level.

Mortimer et al., 1977

"The product of respiration is carbon dioxide, and it is an essential component of the life process. Producing and retaining enough carbon dioxide is as important for longevity as conserving enough heat to allow chemical reactions to occur as needed. Carbon dioxide protects cells in many ways. By bonding to amino groups, it can inhibit the glycation of proteins during oxidative stress. It can limit the formation of free radicals in the blood; inhibition of xanthine oxidase is one mechanism (Shibata et al., 1998). It can reduce inflammation caused by endotoxin/ LPS by lowering the formation of tumor necrosis factor, IL-8, and other inflammation promoters (Shimotakahara et al., 2008). It protects mitochondria (Lavani et al., 2007), maintaining (or even increasing) their ability to respire during stress," writes Dr. Ray Peat.

"The suppression of mitochondrial respiration increases the production of toxic free radicals, and the decreased carbon dioxide makes the proteins more susceptible to attack by free radicals. The presence of carbon dioxide is an indicator of proper mitochondrial respiratory functioning. In every type of tissue, it is the failure to oxidize glucose that produces oxidative stress and cellular damage," Dr. Ray Peat says, and then concludes, "A focus on correcting the respiratory defect would be relevant for all diseases and conditions (including heart disease, diabetes, dementia) involving inflammation and inappropriate excitation, not just for cancer. Carbon dioxide has a stabilizing effect on cells, preserving stem cells, limiting stress, and preventing loss of function."

> *Over the oxygen supply of the body*
> *carbon dioxide spreads its protecting wings.*
> *Friedrich Miescher*

Swiss physiologist, 1885

Without enough oxygen, the electron transport chain becomes jammed with electrons. Consequently, NAD[192] cannot be produced, causing glycolysis to produce lactic acid instead of pyruvate, a necessary component of the Krebs Cycle.

In general, we tend to assume that cancer cells generate energy using glycolysis rather than mitochondrial oxidative phosphorylation and that the mitochondria are dysfunctional. Advances in research techniques have shown the mitochondria in cancer cells to be partially functional across various tumor types. However, different tumor populations have various bioenergetic alterations to meet their high energy requirement meaning the Warburg effect is not consistent across all cancer types.

CO2 has Antioxidant Properties.

Normal arterial levels of CO2 have antioxidant properties. Indeed, a group of Russian microbiologists discovered that "CO2 at a tension close to that observed in the blood (37.0 mm Hg) and high tensions (60- or 146-mm Hg) is a potent inhibitor of the generation of the active oxygen forms (free radicals) by the cells and mitochondria of the human and tissues" (Kogan et al., 1997).

Health state	Type of breathing	Degree	Pulse, beats/min	Breathing frequency/ min	CO2 in alveoli, %	AP, s	CP, s	MP, s
Super-health	Shallow	5	48	3	7,5	16	180	210
		4	50	4	7,4	12	150	190
		3	52	5	7,3	9	120	170
		2	55	6	7,1	7	100	150
		1	57	7	6,8	5	80	120
Normal	Normal	-	60	8	6,5	4	60	90
Disease	Deep	-1	65	10	6,0	3	50	75
		-2	70	12	5,5	2	40	60
		-3	75	15	5,0	-	30	50
		-4	80	20	4,5	-	20	40
		-5	90	26	4,0	-	10	20
		-6	100	30	3,5	-	5	10

As we have seen, arterial hypocapnia (CO_2 deficiency) causes tissue hypoxia that triggers numerous pathological effects. Cell hypoxia is the leading cause of free radical generation and oxidative stress. CO_2 deficiency in the blood is one of the leading causes of hypoxia (low oxygen).

Having an average level of CO2 in the lungs and arterial blood (40 mm Hg or about 5.3% at sea level) is imperative for everyday health. Do modern people have normal CO2 levels? When reading the table below, note that levels of CO2 in the lungs are inversely proportional to minute ventilation rates; in other words, the more air one breaths, the lower the level of alveolar CO2.

Fast Breathing.

Dozens of studies have shown that modern "normal subjects" breathe about 12 L/min at rest, while the medical norm is only 6 L/min. As a result, blood CO2 levels are less than average.

Dr. Artour Rakhimov writes, "The minus 4th and -5th degrees of health in the above chart corresponds to patients whose life is not threatened at the moment, but their main concern is symptoms. People with mild asthma, heart disease, diabetes, initial stages of cancer, and many other chronic disorders are all in this zone. Taking medication is the standard feature for most of these people. As we see from the table, these patients' heart rate varies from 80 to 90 beats per minute. Breathing frequency is between 20 and 26 breaths per minute (the medical norm is 12, while doctor Buteyko's norm is eight breaths per minute at rest). Physical exercise is very hard since even fast walking results in very heavy breathing through the mouth, exhaustion, and worsening symptoms. Complains about fatigue are normal. All these symptoms are often so debilitating that they interfere with normal life and the ability to work, analyze information, or care about others. Living in a chronic state of anxiety because of stress and being preoccupied with one's miserable health is normal. At the same time, efficiency and performance in various areas (science, arts, sports, etc.) are compromised. Sitting in armchairs or soft couches is the most favorite posture."

Dr. Lynne Eldridge and many others have noted most modern adults breathe much faster than what would be considered a healthy respiratory rate. Respiratory rates in cancer and other severely ill patients are usually higher, generally about 20 breaths/min or more. Meaning the general population is driving down oxygen available to cells opening the door to increased cancer incidences. The cells' heavy metal and chemical toxicity further impede oxygen with nutritional deficiencies are a slam dunk that leads to cancer.

Oxygen availability to cells decreases glucose oxidation, whereas oxygen shortage consumes glucose faster to produce ATP via the less efficient anaerobic glycolysis to lactate. This is much of the basis of oxygen therapy in cancer and a full range of other diseases because most chronically ill people, if not all, are having a challenging time with both oxygen and its perfectly mated gas, carbon dioxide.

Perfectly Normal Until its Cancer.

We can take some lessons from our muscles when they worked hard. When the body has plenty of oxygen, pyruvate is shuttled to an aerobic pathway to be broken down for more energy. But when oxygen is limited, the body temporarily converts pyruvate into lactate, allowing glucose breakdown and energy production—to continue. Even in healthy athletic individuals, when we put the muscles to great challenges, oxygen levels fall temporarily, showing us what happens in cells when they are oxygen-starved.

In cancer, the change becomes permanent. Cancer cells will continue with the fermentation of glucose and lactate production, even in the presence of oxygen. However, evidence shows that some cancer cells, especially young cancer cells, can be reverted to normal cells if they have enough oxygen.

The lactic acid in our tissues is a cause of biological problems for many reasons; lactic acid displaces carbon dioxide. The primary features of stress metabolism include increases in stress hormones, lactate, ammonia, free fatty acids, fat synthesis, and decreased carbon dioxide. The lactic acid in the blood can be taken as a sign of defective respiration since glucose breakdown to lactic acid increases to make up for deficient oxidative energy production.

Glucose can be metabolized into pyruvic acid, which changes into carbon dioxide in the presence of oxygen. Without oxygen, pyruvic acid transforms into lactic acid. The decrease in carbon dioxide accompanies increased lactic acid production.

The ability of lactic acid to displace carbon dioxide participates in the blood clotting system. It contributes to disseminated intravascular coagulation and consumption coagulopathy. It increases red cells' tendency to aggregate, forming "blood sludge," making red cells more rigid, increasing blood viscosity, and impairing circulation in the small vessels. (Schmid-Schönbein, 1981; Kobayashi et al., 2001; Martin et al., 2002; Yamazaki et al., 2006.) Lactate and inflammation promote each other in a vicious cycle (Kawauchi et al., 2008).

> *Low thyroid leads to inadequate production of*
> *carbon dioxide and wastage of glucose.*
>
> *Dr. Ray Peat*

Carbon dioxide protects cells in many ways. By bonding to amino groups, it can inhibit the glycation of proteins during oxidative stress. It can limit the formation of free radicals in the blood; inhibition of xanthine oxidase is one mechanism (Shibata et al., 1998). It can reduce inflammation caused by endotoxin/LPS by lowering the formation of tumor necrosis factor, IL-8, and other inflammation promoters (Shimotakahara et al., 2008). CO_2 protects mitochondria (Lavani et al., 2007), maintaining (or even increasing) their ability to respire during stress."

Carbon dioxide stabilizes cells, preserving stem cells, limiting stress, and preventing loss of function. Carbon dioxide can be used to avoid adhesions during abdominal surgery and protect the lungs during mechanical ventilation.

Enough carbon dioxide is essential in preventing an excessive and maladaptive stress response. A deficiency of carbon dioxide (such as can be produced by hyperventilation or by the presence of lactic acid in the blood) decreases cellular energy (as ATP and creatine phosphate) and interferes with the synthesis of proteins (including antibodies) and other cellular materials.

The Oxygen Carbon Dioxide Connection.

Most people have unhealthy breathing habits. They hold their breath or breathe high in the chest or in a shallow, irregular manner. These patterns have been unconsciously adopted, accidentally formed, or emotionally impressed.

Certain "typical" breathing patterns trigger physiological and psychological stress and anxiety reactions. Babies know how to breathe, and you can see their bellies expand as the diaphragm moves down and their bellies swell. Adults breathe more through expanding their chest cavity, and it takes training and discipline to return to more natural breathing patterns that allow for full oxygenation.

A lack of carbon dioxide is harmful though many climate hysterics are running around loudly, proclaiming that we have too much and should put a tax on it. Carbon dioxide is a fundamental component of living matter, as is oxygen and if you disagree, ask plants! When people have bicarbonate deficiencies (acid conditions, which most people develop as they age), they have carbon dioxide deficiencies, translating into oxygen deficiencies.

If a carbon dioxide deficiency continues for a long time, it causes diseases, aging, and cancer because oxygen is not appropriately delivered to tissues. Ancient forms of medicine knew how to establish good breathing habits for increased vitality and freedom from disease. They knew that poor breathing reduces our vitality and opens the door to illness.

Yin Yang of Respiration.

The important thing is the relationship of gases – between carbon dioxide and oxygen. Too much oxygen (relative to the carbon dioxide level), and we feel agitated and jumpy. Too much carbon dioxide (again, relative to the oxygen level), and we feel sluggish, sleepy, and tired.

Most doctors maintain a natural misconception that oxygen and carbon dioxide are antagonistic that a gain of one in the blood necessarily involves a corresponding loss of the other. This is not correct; although each tends to raise the pressure and thus promote the diffusion of the other, the two gases are held and transported in the blood by different means; the hemoglobin in the corpuscles carries oxygen, while carbon dioxide is combined with alkali in the plasma.[193]

A sample of blood may be high in both gases or low in both gases. Under clinical conditions, low oxygen and low carbon dioxide occur together. Therapeutic increase of carbon dioxide, by inhalation of this gas diluted in air, is often an effective means of improving blood and tissue oxygenation.[194]

Wound Healing with Carbon Dioxide and Oxygen.

Look at the profound healing effect of carbon dioxide. The following shows the treatment effects of CO2 medicine for a diabetic foot. Carbon dioxide footbath therapy heals diabetic foot and other ischemic ulcers.[195] This healing was accomplished with sodium bicarbonate baths laced with citric acid, which breaks down the bicarbonate into CO2 microbubbles.

This is before, then one month, and three months after treatment. The only other treatment that comes close to helping a diabetic foot like this is magnesium therapy, which combines beautifully in baths with bicarbonate and CO2 medicine therapies.

We can see and compare the same treatment type with oxygen and see the results are the same. The University of Tennessee Medical School shows what oxygen can do for wound healing.[196] Every cell in our body can recognize and respond to changes in the availability of oxygen. The best example of this is when we climb to high altitudes where the air contains less oxygen. The cells recognize the decrease in oxygen via the bloodstream and react, using the 'hypoxic response' to produce an EPO protein (erythropoietin). In turn, this protein stimulates the body to produce more red blood cells to absorb as much of the reduced oxygen levels as possible. [197]

We can literally force mitochondria to become active again and use the Krebs cycle for energy if we ram enough oxygen into the cells. This process is facilitated when hydrogen is added.

If you put enough oxygen and hydrogen into a cancer cell, it will turn on the Krebs cycle (the mitochondria), which reignites the cell death program. Remember, carbon dioxide is the main product of the Krebs Cycle. So, when carbon dioxide levels go up, we increase our health. Therefore exercise is so important. It is the absolute best way to create lots of CO2! CO2 coming off the mitochondria in ample quantities shows that our energy factories of life are fired up with all burners burning.

Breathing Retraining.

All doctors should know that chronically and even seriously ill people with dangerous acute infections benefit immediately from controlling the quantity of air going into and out of their lungs. With the Frolov, in the space of 20 minutes once or twice a day, one can begin to control critical medical parameters, the most important of which is oxygen delivery to cells and tissues.

"Medical textbooks suggest that adults' normal respiratory rate is only 12 breaths per minute at rest. Older textbooks often provide even smaller values (e.g., 8-10 breaths per minute). Respiratory rates in the sick are usually higher, generally about 20 breaths/min or more," writes Artour Rakhimov.

When we practice breathing retraining, it is like standing on a chariot with four wild horses. We pull back on the reins—limiting the airflow, slowing everything down—we increase electron flow, raising cellular voltage, pH, and oxygenation as well as carbon dioxide levels.

More is Less.

Medical studies have proven that the more we breathe, the less oxygen is provided for the body's vital organs. Does that sound upside down to you? Ideal breathing corresponds to slow, light, and easy abdominal breathing (also called diaphragmatic or belly breathing), which needs to be relearned in most adults. Diaphragmatic breathing allows one to take normal breaths while maximizing the amount of oxygen flowing into the bloodstream.

"Deep breathing" exercises and techniques, to anyone who knows something about breathing, do not suggest in any way that one should over breathe. Deep breathing is just another way of saying belly breathing as opposed to shallow, superficial chest breathing. Deep breathing should be very slow so that one accumulates more CO_2 in the blood. Deep breathing means breathing less air, not more. Some people think it is wrong to call therapeutic breathing 'deep breathing'. If you breathe less and accumulate CO_2, the correct name is 'reduced breathing,'" writes Rakhimov.

When we breathe less—using a breathing device—we directly influence the involuntary (sympathetic nervous system) that regulates blood pressure, heart rate, circulation, digestion, and many other bodily functions. Breath is life, so we can expect to feel alive, vibrant, and healthy if we bring our awareness to our breath and retrain the way we breathe.

When we breathe correctly, we can live more perfectly in health because our breath is the most essential energy source. Hippocrates said, "Air is a pasture of life and the greatest ruler of all." I suppose he knew what ancient oriental philosophers knew—that in the air is "an ocean of energy" ready to be tapped into directly.

We breathe all day, every day, so we might as well do it right. Since a breath is the first and last physical activity we undertake in life, we should consider the importance it deserves in our pursuit of health and relaxation. We can live a long time without food and a couple of days without drinking, but life without breath is measured in minutes. Unfortunately, unless one participates in or teaches yoga, breathing does not get the attention it deserves.

The American Academy of Cardiology says, "Stress can cause shortness of breath or make it worse. Once you start feeling short of breath, it is common to get nervous or anxious. This can make your shortness of breath even worse. Being anxious tightens the muscles that help you breathe; this makes you start to breathe faster. As you get more anxious, your breathing muscles get tired. This causes even more shortness of breath and more anxiety. At this point, you may panic."

Learning to avoid or control stress can help you avoid this cycle. You can learn tips to help you relax and learn breathing techniques to get more air into your lungs.

American Academy of Cardiology

Benefits of Slow Breathing.

1. Breathing detoxifies and releases toxins.
2. Breathing releases tension.
3. Breathing relaxes the mind/body and brings clarity.
4. Breathing relieves emotional problems.
5. Breathing relieves pain.
6. Breathing massages your organs.
7. Breathing increases muscle.
8. Breathing strengthens the immune system.
9. Breathing improves posture.
10. Breathing improves the quality of the blood.
11. Breathing increases digestion and assimilation of food.
12. Breathing improves the nervous system.
13. Breathing strengthens the lungs.
14. Proper breathing makes the heart stronger.
15. Proper breathing assists in weight control.
16. Breathing boosts energy levels and improves stamina.
17. Breathing improves cellular regeneration.
18. Breathing elevates moods.

Even *Readers Digest* got into writing about breathing saying, "What could be more basic than breathing? Inhale, exhale, repeat... right? Not exactly. Western science and medicine focus on breathing as a bodily function integral to survival. Eastern health sciences approach it as nourishment for both body and spirit. The Chinese believe that mindful breathing, or "breath work," has numerous benefits, including improved focus and efficiency, increased positivity, and greater physical and mental energy."

"The body can store many of the things it needs to function, such as vitamins and food, in the form of fat. Oxygen is one item that cannot be stored in sufficient quantities for more than a few minutes. The blood holds about a quart of dissolved oxygen at rest, but it is continually being used by the cells to produce energy. The lungs need to be constantly working to furnish a sufficient supply for various activities."

Your breathing or respiratory rate is defined as the number of breaths a person takes during one minute while at rest. Recent studies suggest that an accurate recording of respiratory rate is significant in predicting severe medical events. Since many factors can affect the results, understanding how to take a precise measurement is essential. While watching the clock, count the number of times you breathe in two minutes. Make three trials and find the average. Divide by two to find the average number of breaths per minute.

The rate should be measured at rest, not after someone has been up and walking about. Being aware that your breaths are being counted can make the results inaccurate, as people often alter the way they breathe if they know it is being monitored. Nurses are skilled at overcoming this problem by discretely counting respiration, watching the number of times your chest rises and falls — often while pretending to take your pulse.

Lung expert Dr. Lynne Eldridge says that "In general, children have faster respiratory rates than adults, and women breathe more often than men. The normal ranges for different age groups are listed below:

Newborn: 30-60 breaths per minute
Infant (1 to 12 months): 30-60 breaths per minute
Toddler (1-2 years): 24-40 breaths per minute
Preschooler (3-5 years): 22-34 breaths per minute
School-age child (6-12 years): 18-30 breaths per minute
Adolescent (13-17 years): 12-16 breaths per minute
Adult: 12-18 breaths per minute (Fast breathing is the 'new' normal)

Dr. Sheldon Saul Hendler writes, "Breathing is unquestionably the single most important thing you do in your life. And breathing right is the single most important thing you can do to improve your life." So, what is the actual difference between our lives and health when we breathe less? You will be astounded by the information that Michael White has put together. 85,000 people filled out his questionnaire on his site yielding the following vital information:

Complete Breaths vs. Diagnosed Conditions

B/K1	d. Anxiety or panic attacks			f. Attention issues			t. High blood pressure			ee. Sleeping disorders			m. Depression			z. Overweight/Obese		
	% of total test takers	% of total test takers with row choice	% of total test takers with column choice	% of total test takers	% of total test takers with row choice	% of total test takers with column choice	% of total test takers	% of total test takers with row choice	% of total test takers with column choice	% of total test takers	% of total test takers with row choice	% of total test takers with column choice	% of total test takers	% of total test takers with row choice	% of total test takers with column choice	% of total test takers	% of total test takers with row choice	% of total test takers with column choice
5-6	1.4	13.8	6.5	0.2	1.8	3.9	1.4	13.8	9.2	0.5	4.6	4.7	0.8	8.3	5.4	1.5	15.6	8.3
7-8	3.0	22.0	14.3	0.5	3.3	9.8	2.2	16.0	14.7	1.2	8.7	12.1	2.3	16.7	14.9	2.6	19.3	14.1
9-11	5.1	22.0	24.2	0.7	3.1	15.7	3.4	15.0	23.3	2.4	10.6	25.2	3.1	13.4	20.2	4.3	18.5	22.8
12-24	9.5	22.0	45.5	2.6	6.1	56.9	6.1	14.0	41.1	4.2	9.6	43.0	7.6	17.6	50.0	9.0	20.8	48.1

Courtesy breathing.com

You should stare at this chart for a while and let its information sink in. You can see that slow breathers have health, and fast breathers just having the toughest time with their bodies and lives. Quick breathers suffer from higher anxiety levels, depression, sleeping disorders, and high blood pressure than slow breathers.

Dr. Fred Muench says, "Once you go below ten breaths a minute, you start to engage the parasympathetic nervous system, which helps the body relax when it has been injured. Slow breathing activates the vagus nerve, the primary cranial nerve, which is associated with a recuperative state." Perhaps more important, slow breathing tends to increase heart-rate variability, reflecting the fluctuation in heartbeats during an activity. "If your heart rate fluctuates 60 to 80 beats per minute, cardiac-wise, that's healthier than someone whose heart rate varies between only 70 and 75 beats per minute," says Muench. "It means your system is not so rigid. Someone like Lance Armstrong has a massive swing in heart-rate variability, whereas an unhealthy or older person has a much smaller one. The way to increase variability is to breathe slowly."

A person who is breathing at four breaths a minute will only breathe about 5,760 times a day. At the "normal" breathing rate of eight breaths, a minute counts double to 11,520 breaths a day. At 16, which is still slow for many ill people, that rate reaches 23,000 breaths a day. At 25 breaths a minute, we are clipping along at 36,000 breaths a day, which is a far cry above a normal rate.

Dr. Buteyko found that virtually all sick people (asthma, bronchitis, heart disease, diabetes, cancer, etc.) have accelerated respiratory patterns. During rapid breathing, carbon dioxide becomes deficient, oxygen delivery to the cells is reduced, breath-holding time is reduced, and the natural automatic pause is absent in each breath.

After thirty years of studying over 5,000 patients in the Framingham studies, doctors from the Boston University School of Medicine said they could predict long-term and short-term mortality based on people's breathing capacity. Dr. William Kannel said a person's vital breathing capacity can "Pick out people who are going to die 10, 20, or 30 years from now."

In their book *Perfect Breathing,* Al Lee and Don Campbell say, "The impact of the breath extends into every aspect of life and shows itself at the root of human function. Ancient teachers, sages, yogis, and martial artists discovered its power and developed disciplines around it with yoga, Qigong, and karate, among so many other practices. Understanding the breath means understanding the human machine and how each breath can develop and control the body. Breathing forms the foundations of meditation, contemplative thought, and prayer. Still, it is also informing science and medicine, as conscious breathing proves its mettle as a tool to fight stress, build up immunity to disease, and heal the body in many ways. What is a perfect breath? Far from being some noble yet unreachable goal that takes years of rigorous practice to master, a perfect breath is any breath you take for which you are completely mindful and aware. In the space of that one simple breath, great things can be accomplished. Perfect breathing is attainable and within easy reach. Your very next breath can be a perfect breath."

Most modern people suffer from breathing problems. The common issues include chest breathing (as opposed to abdominal diaphragmatic), mouth breathing, and hyperventilation (breathing too fast), all of which reduce oxygen levels in body cells and promote chronic diseases. The bottom line is, the quicker we breathe, the sicker we become. Breathing too fast will end one in a casket if one is not careful, and indeed, life will be lived with pain and disease unless one gets control.

Dr. Nerina Ramlakhan writes, "I noticed that over 75% of the clients (not patients - these were 'well' corporate employees) were breathing sub-optimally in ways that would make them stressed, chronically exhausted, overweight, and insomniac. It is imperative to our health to get our breathing right. My advice is simple - just for 21 days, become even slightly obsessed with your breathing. Notice it five times a day: first thing in the morning before you get out of bed, last thing at night as you drift off to sleep, and then find three other times during the day.

Put your feet on the ground, drop your breathing into your belly, exhale long and inhale fat. Breathe well consciously so later you can breathe well unconsciously."

Warning: Depending on the severity and type of the condition, patients can worsen their health if they go into intensive breathing sessions too aggressively. Some critically ill patients can develop even higher blood pressure, panic attacks, and migraine headaches from aggressive and rapid breathing changes.

Diaphragmatic Breathing.

When you take a truly deep breath, you are expanding the lungs, pressing down the diaphragm, and causing your abdomen to expand as your lungs fill with air. This is not only wonderful for reducing tension, but research has shown that it may also help with diseases that inhibit breathing, like emphysema.

Diaphragmatic breathing effectively calms us down. It also makes sure that you take in lots of oxygen. If you are not sure, you breathe deeply enough, try lying down, and put a magazine on your stomach. Make sure you expel all your air, exhaling completely, and then slowly raise the magazine as you inhale. Inhale for five nice, long counts. Exhale the same way, counting until the magazine goes down. You can also use your hand instead of a magazine.

Breathing retraining has a lot to offer anxious hyper-tense patients. Worried people are experiencing the sympathetic nervous system (the fight or flight system) over-arousal. Slow breathing reduces the sympathetic nervous system over-arousal and increases parasympathetic nervous system activity – the relax, recuperate, regenerate system –which calms people down.

Conscious Breathing and its Effect on the Heart.

According to yoga, energy flows more freely through the heart when we breathe into it and focus our attention on that area – energy flows where attention goes. Breathing is linked to and directly affects the heart. The regular practice of diaphragmatic breathing significantly improves heart rate variability and coherence. All dynamics of the heart are improved when we breathe correctly. The more a person improves their heart rate variability (HRV), the healthier they become. This is good for ischemic heart patients who have diabetes.[198] Heart rate variability is indispensable in distinguishing healthy subjects from patients with cardiovascular disease.

Heart rate variability (HRV) is an indicator of cardiac autonomic control. Yogic, or what is known as belly breathing (deep abdominal breathing vs. shallow rib cage breathing), modifies the autonomic status by increasing sympathetic activity (reduced vagal activity). It is uncanny how accurately low HRV scores and trends align with illness.

Ancient yoga breathing techniques regulate the heart, stimulate and increase vital energy, strengthen internal organs, and regenerate and rejuvenate the body. Through breathing consciously, we can achieve the optimal functioning of the endocrine, nervous, digestive, and other bodily systems and gain mental and physical stability.

When your exhale is even a few counts longer than your inhale, the vagus nerve sends a signal to your brain to turn up your parasympathetic nervous system and turn down your sympathetic nervous system. The sympathetic system commands our fight or flight response, and when it is stimulated, it increases heart rate and breathing and increases stress hormones like cortisol. The parasympathetic system, on the other hand, controls your rest-relax-and-digest response. When the parasympathetic system is dominant, your breathing slows, your heart rate drops, your blood pressure lowers as the blood vessels relax, and your body is put into a state of calm and healing.

Mantak Chia wrote, "For thousands of years, Taoist masters have taught natural breathing. We can improve the functioning and efficiency of our heart, lungs, and other internal organs and systems. We can help balance our emotions. We can transform our stress and negativity into the energy that we can use for self-healing and self-development. And we are better able to extract and absorb the energy we need for spiritual growth and independence." Breathing correctly is essential for living longer, and it helps us maintain positive emotions and help keep our performance at its best in everyday activity.

We all breathe, all day, every day, so we might as well do it right. As soon as we pay attention to our breathing, it immediately changes, which is the whole point. Breathing retraining entails bringing our awareness to our breath and treating with respect something essential to maintaining our lives.

Crystal Tatum says, "Breathe. Just breathe. It is so simple; it cannot possibly help, can it? What do you mean, just breathe? Of course, I am breathing! What a dumb thing to say. I have the good fortune of being friends with many highly evolved folks who know a thing or two about helping the not-so-highly evolved, such as myself. But when one of those friends said to me one day, "Don't forget to breathe," I couldn't help but cock an eyebrow and give her a "What the heck are you talking about?" look. She told me I was holding my breath. I thought she was nuts, but the next time I found myself angst-ridden, I took notice of my body and realized she was right. Since then, I have noticed that I tend to do that when I'm highly stressed or anxious. I clench my jaw and hold my breath, taking only the shallowest inhalations when necessary. This response only heightens my stress and keeps me on edge. I've learned a few breathing techniques since then that really does ease my tension."

Dennis Lewis, the author of the *Tao of Breathing,* wrote, "In 1990, I found myself physically, emotionally, and spiritually exhausted, with constant, sharp pain on the right side of my rib cage. When Gilles Marin first put his hands into my belly and began to massage my inner organs and tissues, and when he began to ask me to breathe into parts of myself that I had never experienced through my breath, I had no idea of the incredible journey of discovery that I was beginning. Though the physical pain disappeared after several sessions, and though I began to feel more alive, a deeper, psychic pain began to emerge—the pain of recognizing that despite all my efforts over many years toward self-knowledge and self-transformation. I had managed to open myself to only a small portion of the vast scale of the physical, emotional, and spiritual energies available to us at every moment. As Gilles continued working on me, and as my breath began to penetrate deeper into myself, I began to sense layer after layer of tension, anger, fear, and sadness resonating in my abdomen below the level of my so-called waking consciousness and consuming the energies I needed not only for health but also for a real engagement with life. And this deepening sensation at the very center of my being, painful as it was, brought with it an opening, not only in the tissues of my belly but also in my most intimate attitudes toward myself, a welcoming of hitherto unconscious fragments of myself into a new sense of discovery."

PEMF Increases
Oxygen Utilization.

Hydrogen is not the only thing in the universe that will make things burn brighter. In 2003 NASA's Dr. Thomas Goodwin found that a 10 Hz pulsed electromagnetic field caused neural tissue regeneration at four times baseline. Both 5 and 15 Hz provided two times neural tissue regeneration with a perfect bell-shaped curve around 10 Hz. In 1989 Dr. David Hood found chronic (35+ days) 10 Hz stimulation at 10 hours per day increased two critical enzymes of cell respiration by a factor of three.

There was a movie about scientists looking for the ultimate hydrogen reaction; they searched for the perfect frequency to inject into a hydrogen plasma field. In real life, these frequencies are known as the Schumann Waves, the earth frequency.

Mitochondria, under stimulation by pulsed electromagnetic fields (PEMF), synthesize more ATP from the oxygen we breathe at these low-frequency ranges. Cells burn oxygen more efficiently, drawing oxygen from the blood more slowly, producing more energy and less waste from every breath, resulting in an abundance of cellular energy causing these profound regenerative effects.

This extra energy is available at night for repair, hormone synthesis, memory consolidation, and immune support, and during the day for physical and mental performance enhancement (ergogenic effects). More energy from every breath you take is like the claims made when using singlet oxygen therapy.

Dr. Dominic D'Agostino, a researcher and assistant professor with the University of South Florida Morsani College of Medicine, said, "cancer is starved" by eating a restricted diet in carbohydrates but high in certain fats. The patient then receives hyperbaric chamber treatments, in which oxygen has a further toxic effect on the cancer cells, explaining a possible one-two punch to knock out cancer.

D'Agostino began research nine years ago involving metabolic therapy and hyperbaric oxygen to help Navy SEAL divers avoid seizures from oxygen toxicity. A 10-year-old boy with a cancerous brain tumor who had already received a battery of traditional conventional radiation therapy has gotten positive results from this. The youth responded "remarkably" to the combination of diet and hyperbaric treatment.

Low-level electromagnetic fields are known and used to halt cancer cell growth. Costa *et al. (2011)* reported surprising clinical benefits from using the specific EMF signals to treat advanced hepatocellular carcinoma, stabilizing the disease, and producing partial responses up to 58 months in a patient's subset. Now Zimmerman *et al.* have examined the growth rate of human tumor cell lines from liver and breast cancers along with normal cells from those tissues exposed to AM-EMF. The reduced growth rate was observed for tumor cells exposed to tissue-specific AM-EMF, but no change in growth rate in normal cells derived from the same tissue type or in tumor or normal cells from the other tissue type.[199]

In plain language, low-frequency pulses create a brief, intense voltage around each cell. The mitochondria within the cell grab some of this energy. This, in turn, makes the cell more efficient at producing ATP and delivering oxygen throughout the body. PEMF therapy supports metabolism and increases the blood flow by dilating micro capillaries throughout the body, allowing all cells to breathe and function better.

Darkfield microscopy shows that clustering in the erythrocytes can be cleared with PEMF. This leads to improved viscosity of the blood, improved blood flow, enlargement of the surface area, increased oxygen levels, and reduced risk of thrombosis. Thermo-graphic measuring charts show increased circulation after exposure to PEMF, leading to better nutrition and cells' rejuvenation.

All biological processes and, in particular, every cell's metabolism is substantially based on electromagnetic energy. Only an organism sufficiently supplied with energy can control the self-regulating mechanisms and powers of regeneration and healing. One of the common constituents of all cells are ions. Ions are positively and negatively charged particles that conduct electromagnetic pulses from within the cell. The electromagnetic pulses allow the cell to function. PEMFs affect ion flow through specific cell membrane channels (like sodium, potassium, and calcium), positively involving these enzymes. Without ions, a cell cannot live. Without sufficient energy fields, cells do not function at 100%.

Diseased or damaged cells have altered rest potential. If the ions (electrically charged particles surrounding the cells) move into an area of pulsating magnetic fields, they will be influenced by the pulsation rhythm. The rest potential of the cell is proportional to the ion exchange occurring at the cell membrane. The ion exchange is also responsible for the oxygen utilization of the cell. Pulsating magnetic fields can dramatically influence the ion exchange at the cellular and subcellular levels, thereby significantly improving the oxygen utilization of diseased or damaged tissues. The deterioration of oxygen utilization is a problem in several medical branches, especially in delayed healing and arthritis of joints. Poor oxygen utilization = increased oxidative stress that results in worse oxygen utilization.

All atoms, chemicals, and cells produce electromagnetic fields (EMFs). Every organ in the body produces its signature bio-electromagnetic field. Science has proven that our bodies project their magnetic fields and that all 70 trillion cells in the body communicate via electromagnetic frequencies. Nothing happens in the body without an electromagnetic exchange. When the electromagnetic activity of the body ceases, life ceases. When we increase electromagnetic energies, we increase life and promote healing.

Earth Pulse.

I am happy using is the EarthPulse machine that puts out earth & Schumann frequencies below the 14hz area. It is a low-cost but potent machine designed to use all night long while sleeping. Unlike more expensive pulsed electromagnetic field systems, these long nightly sessions are where the magic of the EarthPulse occurs. EarthPulse has numerous reports of waking saturated blood oxygen increasing levels by 5% in just a few days.

EarthPulse allows magnetic field supplementation through footwear, clothing, or at nighttime through your mattress or pillow. PEMF reduces inflammation via several mechanisms, including restoration of cell membrane homeostasis, attenuating pro-inflammatory cytokine Interleukin-1beta (IL-1β) by 10-fold, by lowering expression of major pro-inflammatory genes and increasing expression of anti-inflammatory genes.

The FlexPulse is another excellent system at the lower end of what PEMF equipment can cost you. The EarthPulse is best for all-night treatment and the FlexPulse when you need to direct the frequency waves at a specific area.

Doctors in Brazil have found that pulsed electromagnetic field exposure relieves microwave exposure by inducing Faraday currents. Electromagnetic fields are recognized as hazards that affect testicular function by generating reactive oxygen species and reducing androgen's bioavailability to maturing spermatozoa. Thus, microwave exposure adversely affects male fertility, whereas pulsed electromagnetic field therapy is a non-invasive, simple technique used as a scavenger agent to combat oxidative stress.[200]

PEMF Treats Cancer and Improves Oxygen Delivery.

Pulsed Magnetic Field Therapy (PEMF) is FDA approved to promote bone union healing. It has been used in Europe for over 20 years in 400,000 sessions with individuals with cancer, migraines, sports-related injuries, wound healing, and other pain syndromes. PEMF-based anticancer strategies represent a new therapeutic approach to treat breast cancer without affecting normal tissues in a manner that is non-invasive and can be combined with existing anti-cancer treatments.

Costa et al. (2011) reported surprising clinical benefits from using the specific EMF signals to treat advanced hepatocellular carcinoma, stabilizing the disease, and producing partial responses up to 58 months in a patient's subset. Now Zimmerman et al. have examined the growth rate of human tumor cell lines from liver and breast cancers along with normal cells from those tissues exposed to AM-EMF. The reduced growth rate was observed for tumor cells exposed to tissue-specific AM-EMF, but no change in growth rate in normal cells derived from the same tissue type or in tumor or normal cells from the other tissue type.

In layman's terms, low-frequency pulses create a brief, intense voltage around each cell. The mitochondria within the cell grab some of this energy. This, in turn, makes the cell more efficient at producing ATP and delivering oxygen throughout the body. PEMF therapy supports metabolism and increases the blood flow by dilating micro capillaries throughout the body, allowing all cells to breathe and function better.

All biological processes, including the metabolism of every single cell, are based on electromagnetic energy. Only an organism sufficiently supplied with energy can control the self-regulating mechanisms and powers of regeneration and healing. One of the common constituents of all cells are ions. Ions are positively and negatively charged particles that conduct electromagnetic pulses from within the cell. The electromagnetic pulses allow the cell to function. PEMFs affect ion flow through specific cell membrane channels (like those for sodium, potassium, and calcium), which positively affect these enzymes. Without ions, a cell cannot live. Without sufficient energy fields, cells do not function at 100%.

Diseased or damaged cells have altered rest potential. If the ions (electrically charged particles surrounding the cells) move into an area of pulsating magnetic fields, they will be influenced by the pulsation rhythm. The rest potential of the cell is proportional to the ion exchange occurring at the cell membrane. The ion exchange is also responsible for the oxygen utilization of the cell. Pulsating magnetic fields can dramatically influence the ion exchange at the cellular and subcellular levels, thereby significantly improving the oxygen utilization of diseased or damaged tissues. The deterioration of oxygen utilization is a problem in several medical branches, especially in delayed healing and arthritis of joints. Low oxygen utilization = increased oxidative stress that results in worse oxygen utilization.

All atoms, chemicals, and cells produce electromagnetic fields (EMFs). Every organ in the body produces its signature bio-electromagnetic field. Science has proven that our bodies project their magnetic fields and that all seventy trillion cells in the body communicate via electromagnetic frequencies. Nothing happens in the body without an electromagnetic exchange. When the electromagnetic activity of the body ceases, life ceases. When we increase electromagnetic energies, we increase life and promote healing.

Another way PEMFs increase oxygen is in their power to reduce chronic, damaging inflammation. PEMFs can induce the appropriate death of aged, chronic T lymphocytes through T cell membranes and key enzymes in cells. The elimination of T cells can minimize the unwanted effects of inflammation, accelerate healing, and reduce the risk of chronic inflammatory diseases.

Alternative Medicine.

How do we practice alternative medicine? What alternative medicine exactly is and isn't is the real question—what is the best way to practice medicine? There must be a right way and a wrong way to practice medicine. The concept of malpractice sustains this belief. Medical malpractice kills 500 times more Americans than accidental gun deaths do. An eight-year study from Johns Hopkins found that there are at least 250,000 deaths due to malpractice in America each year. Other studies have found such deaths surpass 400,000 per year. We also know that over 100,000 Americans die each year from properly prescribed medicines.

As presently defined, alternative medicine is the practice of anything other than mainstream medicine, which means pharmaceutical medicine and other foul things like nuclear medicine and vaccine medicine. I created Natural Allopathic Medicine not as an alternative to modern medicine but as a replacement for it. It makes sense to replace a vast killing machine with something safe and effective.

Wikipedia lumps together, "Alternative medicine, fringe medicine, pseudo-medicine or simply questionable medicine is the use and promotion of practices which are unproven, disproven, impossible to prove, or excessively harmful concerning their effect — in the attempt to achieve the healing effects of medicine." Yet we know most modern medicine is dangerous, and rarely does the word healing go with western medicine. We know that hordes of people are laid low, disabled, made miserable, or outright killed by our poison-happy western medical-industrial complex.

According to the National Center for Complementary and Integrative Health, over 30 percent of Americans use some form of non-conventional medicine. Others say that about 40% of adults in the United States say they use some form of alternative medicine. Complementary and alternative medicine, sometimes referred to as CAM, is an umbrella term for many treatments that fall outside conventional Western approaches. Some have been well-studied and proven to be effective; others have not.

Your Choice.

A complementary therapy means you can use it alongside your conventional medical treatment. It may help you to feel better and cope better with your cancer and treatment. An alternative therapy, on the other hand, is generally used instead of conventional medical treatment. Here in Brazil, the government completely embraces most forms of alternative and commentary medicine.

It is hard to believe when someone writes that all conventional cancer treatments, such as chemotherapy and radiotherapy, must undergo rigorous testing by law to prove that they work. There is no such proof, and it is no secret that toxic cancer treatments do not end well.

Pharmaceutical cheerleaders like to say that most alternative therapies have not been through such testing, and there is no scientific evidence that they work. And they love to say that some alternative therapy types may not be completely safe and could cause harmful side effects. They hide from the public as much as possible how many children get hurt from vaccines and how many people are legally murdered.

One must leave it up to each person to decide what to believe and trust, though some are overly aggressive and force people to choose. Vaccines are forced into babies, and we all know how oncologists feel about alternatives to chemo and radiation treatments. However, the truth is that alternative medical practices are gaining traction in the U.S. and worldwide as modern medicine's reputation is dragged through the mud.

One of the main points about alternative medicine is that it is often a mosaic of many different practices and viewpoints but not tied together into a comprehensive whole. Professionals like acupuncturists, chiropractors, homeopaths, and naturopaths offer comprehensive approaches, but differences in practice and theory prevail.

Whole medical systems involve complete systems of theory and practice that have evolved independently from or parallel to allopathic (conventional) medicine. Many are traditional systems that are practiced by diverse cultures throughout the world.

There is a right way, a universal medical path that eventually everyone can agree on and follow. Something so tangible and fundamental that in a thousand or even a million years, its truth will remain self-evident. What is so universal and constant that such a thing could be said?

Hydrogen in the sun being burnt up as fuel is one such thing. Oxygen being necessary for life is another. That carbon dioxide controls the very foundation of health is yet another. The list goes on that will never change for healthy cell life. One cannot live a good long life if one is deficient in essential minerals like magnesium, iodine, selenium, sulfur, and bicarbonates in the blood. Does it not make sense to lay the foundation of sane medical practice on such absolute necessities of biological life?

Conclusion.

We only must worry about gasoline, oil, water, and electricity to maintain function with a car. We must take care of about fifteen main factors in humans instead of just these four for your vehicle. However, even with fifteen principle medicines and therapeutic processes, it is not that much harder to understand and practice Natural Allopathic Medicine than pump gas or ride a bicycle.

Take a deep breath, go for a walk, listen to your body. A minor physical discomfort usually tells you where to put your attention. In this case, help your body with inexpensive nutritional medicine. If your body is already screaming for help, be even more gentle and avoid additional stress from aggressive medical practices. Hydrogen medicine, combined with oxygen and CO2, is a gentle but effective way to regain health and vitality for you and your family.

References

1   Recent Advances in Hydrogen Research as a Therapeutic Medical Gas Article·Literature Review in Free Radical Research 44(9):971-82 · September 2010

2   Med Gas Res. 2011; 1: 18. Molecular hydrogen protects chondrocytes from oxidative stress and indirectly alters gene expressions through reducing peroxynitrite derived from nitric oxide

3   Med Gas Res. 2011; 1: 11. Effects of drinking hydrogen-rich water on the quality of life of patients treated with radiotherapy for liver tumors.

4   Int J Radiat Biol. 2015 Jan;91(1) Ionizing radiation-induced oxidative stress, epigenetic changes and genomic instability: the pivotal role of mitochondria.

5   Cancer Lett. 2012 Dec 31; 327(0): 48–60. Ionizing radiation-induced metabolic oxidative stress and prolonged cell injury

6   "At the end of the 19th century, scientists Bohr and Verigo discovered what seemed a strange law: A decreased level of carbon dioxide in the blood leads to decreased oxygen supply to the cells in the body, including the brain, heart, kidneys, etc. Carbon dioxide (CO2) was found to be responsible for the bond between oxygen and haemoglobin." – Dr. Alina Vasiljeva and Dr. David Nias. The Bohr Effect was first introduced, describing the oxygen-binding affinity of the hemoglobin as inversely proportional to pH and the concentration of carbon dioxide. In practice, this means that if the carbon dioxide concentration increases somewhere in the body, the hemoglobin molecule will bind to oxygen with lower affinity; therefore, a larger amount of oxygen is released to the area concerned.

7   Inhaled medical gases: more to breathe than oxygen.Respir Care. 2011 Sep;56(9):1341-57; discussion 1357-9. doi: 10.4187/respcare.01442.

8   https://en.wikipedia.org/wiki/CD34

9   Vol 7, No 4 (August 2018) / Therapeutic potential of molecular hydrogen in ovarian cancer

10   Hyperbaric Hydrogen Therapy: A Possible Treatment for Cancer Author(s): Malcolm Dole, F. Ray Wilson, William P. Fife Source: Science, New Series, Vol. 190, No. 4210 (Oct. 10, 1975), pp. 152-154

11   Roberts BJ, Fife WP, Corbett TH, Schabel Jr FM. Response of five established solid transplantable mouse tumors and one mouse leukemia to hyperbaric hydrogen. Cancer Treat Rep. 1978;62(7):1077–9.

12   Gharib B, Hanna S, Abdallahi OM, Lepidi H, Gardette B, De Reggi M. Anti-inflammatory properties of molecular hydrogen: investigation on parasite-induced liver inflammation. C R Acad Sci III. 2001;324(8):719–24.

13   Med Gas Res. Jul-Sep 2019;9(3):115-121."Real world survey" of hydrogen-controlled cancer: a follow-up report of 82 advanced cancer patients. Ji-Bing Chen, Xiao-Feng Kong, You-Yong Lv, Shu-Cun Qin, Xue-Jun Sun, Feng Mu, Tian-Yu Lu, Ke-Cheng Xu.

14   Hydrogen acts as a therapeutic antioxidant by selectively reducing cytotoxic oxygen radicals. Ohsawa I, Ishikawa M, Takahashi K, Watanabe M, Nishimaki K, Yamagata K, Katsura K, Katayama Y, Asoh S, Ohta S. Nat Med. 2007 Jun; 13(6):688-94.

15   Exp Oncol 2009 31, 3, 156–162 PLATINUM NANOCOLLOID-SUPPLEMENTED HYDROGENDISSOLVED WATER INHIBITS GROWTH OF HUMAN TONGUE CARCINOMA CELLS PREFERENTIALLY OVER NORMAL CELLS

16   Molecular Hydrogen as an Emerging Therapeutic Medical Gas for Neurodegenerative and Other DiseasesOxid Med Cell Longev. 2012; 2012: 353152. Published online 2012 Jun 8.doi: 10.1155/2012/353152

17   Med Gas Res. 2011; 1: 18. Molecular hydrogen protects chondrocytes from oxidative stress and indirectly alters gene expressions through reducing peroxynitrite derived from nitric oxide

18   Expression of Metalloproteinases MMP-2 and MMP-9 in Sentinel Lymph Node and Serum of Patients with Metastatic and Non-Metastatic Breast Cancer

19   Cytotechnology. 2012 May; 64(3): 357–371. Suppressive effects of electrochemically reduced water on matrix metalloproteinase-2 activities and in vitro invasion of human fibrosarcoma HT1080 cells

20   PeerJ. 2015; 3: e859. Hydrogen–water enhances 5-fluorouracil-induced inhibition of colon cancer

21   Int J Biol Sci. 2011; 7(3): 297–300. Hydrogen Protects Mice from Radiation Induced Thymic Lymphoma in BALB/c Mice

22   Cancer Chemotherapy and Pharmacology. September 2009, 64:753. Molecular hydrogen alleviates nephrotoxicity induced by an anti-cancer drug cisplatin without compromising anti-tumor activity in mice.

23   Med Gas Res. 2011; 1: 11. Molecular hydrogen alleviates nephrotoxicity induced by an anti-cancer drug cisplatin without compromising anti-tumor activity in mice. Published 2008 in Cancer Chemotherapy and Pharmacology

24   Med Gas Res. 2020 Apr-Jun; 10(2): 75–80. Ji-Bing Chen,[1,2] Xiao-Feng Kong, Feng Mu, Tian-Yu Lu, You-Yong Lu, and Ke-Cheng Xu.

25   Biosci Rep. 2020 Apr 30;40(4):BSR20192761. Hydrogen gas represses the progression of lung cancer via down-regulating CD47 Jinghong Meng, Leyuan Liu, Dongchang Wang, Zhenfeng Yan, Gang Chen

26   Biomed Pharmacother. 2018 Aug;104:788-797. Hydrogen gas inhibits lung cancer progression through targeting SMC3. Dongchang Wang Lifei Wang Yu Zhang Yunxia Zhao, Gang Chen

27   Front. Oncol., 06 August 2019 | https://doi.org/10.3389/fonc.2019.00696 Hydrogen Gas in Cancer Treatment. Sai Li, Rongrong Liao, Xiaoyan Sheng, Xiaojun Luo, Xin Zhang, Xiaomin Wen, Jin Zhou and Kang Peng

28   Dole M, Wilson FR, Fife WP. Hyperbaric hydrogen therapy: a possible treatment for cancer. *Science.* (1975) 190:152–4. doi: 10.1126/science.1166304

29   https://www.youtube.com/watch?v=gyPuP9GMdIw

30   https://www.forbes.com/sites/startswithabang/2017/09/05/the-suns-energy-doesnt-come-from-fusing-hydrogen-into-helium-mostly/#241ce2b270f9

31   https://www.youtube.com/watch?v=SrgQ65UsbZ0

32   PLoS One. 2017; 12(3): e0173645. Hydrogen gas alleviates oxygen toxicity by reducing hydroxyl radical levels in PC12 cells

33   Molecular hydrogen increases resilience to stress in mice. Qiang Gao, Han Song, Xiao-ting Wang, Ying Liang, Yan-jie Xi, Yuan Gao, Qing-jun Guo, Tyler LeBaron, Yi-xiao Luo, Shuang-cheng Li, Xi Yin, Hai-shui Shi & Yu-xia Ma Scientific Reports 7, Article number: 9625 (2017) doi:10.1038/s41598-017-10362-6

34   The Clinical Application of Hydrogen as a Medical Treatment. Acta Med. Okayama, 2016 Vol. 70, No. 5, pp. 331-337. Okayama University Medical School.

35   Brain Research. Volume 1328, 30 April 2010, Pages 152-161

36   Curr Pharm Des. 2013 Oct; 19: 6375–6381. Molecular Hydrogen: New Antioxidant and Anti-inflammatory Therapy for Rheumatoid Arthritis and Related Diseases

37   Beneficial biological effects and the underlying mechanisms of molecular hydrogen - comprehensive review of 321 original articles - Medical Gas Research2015;5:12

38   https://www.gasworld.com/h2-inhalation-research-shows-promise-in-japanese-hospitals/2010537.article

39   Hydrogen Gas Inhalation Treatment in Acute Cerebral Infarction: A Randomized Controlled Clinical Study on Safety and Neuroprotection. http://www.sciencedirect.com/science/article/pii/S105230571730294X

40   Wang R, Wu J, Chen Z et al. Post conditioning with inhaled hydrogen promotes survival of retinal ganglion cells in a rat model of retinal ischemia/reperfusion injury. Brain Res. 2016 Feb 1;1632:82-90.

41   Protective effects of hydrogen-rich saline on monocrotaline-induced pulmonary hypertension in a rat model
Respiratory Research. https://doi.org/10.1186/1465-9921-12-26

42   Kishimoto Y, Kato T, Ito M, Azuma Y, Fukasawa Y, Ohno K, et al. Hydrogen ameliorates pulmonary hypertension in rats by anti-inflammatory and anti-oxidative effects. J Thorac Cardiovasc Surg. 2015;150:645-54.e3.

43   Hydrogen gas reduces chronic intermittent hypoxia induced hypertension by inhibiting sympathetic nerve activity and increasing vasodilator responses via the antioxidation. 27 September 2018 https://doi.org/10.1002/jcb.27684

44   Free Radical Biology and Medicine. Volume 40, Issue 3, 1 February 2006, Pages 398-406
The role of oxidative stress in adult critical care

45   Crit Care. 2006; 10(5): R146. Oxidative stress is increased in critically ill patients according to antioxidant vitamins intake, independent of severity: a cohort study

46   Acute Medicine and Surgery. Promising novel therapy with hydrogen gas for emergency and critical care medicine
24 October 2017. http://onlinelibrary.wiley.com/doi/10.1002/ams2.320/full

47   Hayashida K, Sano M, Kamimura N et al. H2 gas improves functional outcome after cardiac arrest to an extent comparable to therapeutic hypothermia in a rat model. J. Am. Heart Assoc. 2012; 1: e003459.

48   Acute Medicine and Surgery. Promising novel therapy with hydrogen gas for emergency and critical care medicine
24 October 2017. http://onlinelibrary.wiley.com/doi/10.1002/ams2.320/full

49   Perioperative coronary artery spasm in off-pump coronary artery bypass grafting and its possible relation with perioperative hypomagnesemia. Ann Thorac Cardiovasc Surg. 2006 Feb;12(1):32-6.

50   https://pilotonline.com/news/local/health/article_7a3063e5-24cf-56c1-b25c-142731604196.html

51   Ohsawa I, Ishikawa M, Takahashi K et al. Hydrogen acts as a therapeutic antioxidant by selectively reducing cytotoxic oxygen radicals. Nat. Med. 2007; 13: 688–94.

52   Oxidative Medicine and Cellular Longevity. Volume 2016 (2016), Molecular Hydrogen Therapy Ameliorates Organ Damage Induced by Sepsis

53   International Journal of Food Microbiology. Volume 109, Issues 1-2, 25 May 2006, Pages 160-163. Virucidal efficacy of sodium bicarbonate on a food contact surface against feline calicivirus, a norovirus surrogate Yashpal S. Malik and Sagar M. Goyal. Department of Veterinary Population Medicine, College of Veterinary Medicine, University of Minnesota. The

virucidal efficacy of sodium bicarbonate was enhanced when it was used in combination with aldehydes or hydrogen peroxide.

54 Gharib B, Hanna S, Abdallahi OM, Lepidi H, Gardette B, De Reggi M. Anti-inflammatory properties of molecular hydrogen: investigation on parasite-induced liver inflammation. C R Acad Sci III. 2001;324(8):719–24.

55 Hydrogen acts as a therapeutic antioxidant by selectively reducing cytotoxic oxygen radicals.
Ohsawa I, Ishikawa M, Takahashi K, Watanabe M, Nishimaki K, Yamagata K, Katsura K, Katayama Y, Asoh S, Ohta S. Nat Med. 2007 Jun; 13(6):688-94.

56 Cancer Lett. 2012 Dec 31; 327(0): 48–60. Ionizing radiation-induced metabolic oxidative stress and prolonged cell injury

57 International Journal of Cell Biology. Volume 2012 (2012), Article ID 683897, 16 pages. Electromagnetic Fields, Oxidative Stress, and Neurodegeneration

58 Integr Cancer Ther. 2004 Dec;3(4):294-300. Chemotherapy-associated oxidative stress: impact on chemotherapeutic effectiveness.

59 Drug Metabolism and Oxidative Stress: Cellular Mechanism and New Therapeutic Insights
Sharmistha Banerjee, Jyotirmoy Ghosh and Parames C Sil. Division of Molecular Medicine

60 Molecular Hydrogen Alleviates Cellular Senescence in Endothelial Cells. Circulation Journal. Circ J 2016; 80: 2037–2046

61 Trends Biochem Sci. 2002 Jul;27(7):339-44. Oxidative stress shortens telomeres.

62 Med Gas Res. 2011 Oct 3;1(1):24. Open-label trial and randomized, double-blind, placebo-controlled, crossover trial of hydrogen-enriched water for mitochondrial and inflammatory myopathies.

63 Pharmacology & Therapeutics. Volume 144, Issue 1, October 2014, Pages 1-11. Molecular hydrogen as a preventive and therapeutic medical gas: initiation, development and potential of hydrogen medicine

64 A. Mangerich, C. G. Knutson, N. M. Parry, S. Muthupalani, W. Ye, E. Prestwich, L. Cui, J. L. McFaline, M. Mobley, Z. Ge, K. Taghizadeh, J. S. Wishnok, G. N. Wogan, J. G. Fox, S. R. Tannenbaum, P. C. Dedon. PNAS Plus: Infection-induced colitis in mice causes dynamic and tissue-specific changes in stress response and DNA damage leading to colon cancer. Proceedings of the National Academy of Sciences, 2012; DOI: 10.1073/pnas.1207829109

65 http://www.psr.org/environment-and-health/confronting-toxics/heavy-metals/

66 Toxic metals and breast cancer; http://www.townsendletter.com/AugSept2007/toxicmetalbreastcancer0807.htm

67   Hydrogen gas improves survival rate and organ damage in zymosan-induced generalized inflammation model. Xie K, Yu Y, Zhang Z, Liu W, Pei Y, Xiong L, Hou L, Wang G. Shock. 2010 Nov; 34(5):495-501.

68   https://www.hindawi.com/journals/omcl/2015/248529/

69   Med Gas Res. 2013 Jun 6;3(1):11. doi: 10.1186/2045-9912-3-11. Molecular hydrogen: an overview of its neurobiological effects and therapeutic potential for bipolar disorder and schizophrenia.

70   Chen CH, Manaenko A, Zhan Y, Liu WW, Ostrowki RP, et al. (2010) Hydrogen gas reduced acute hyperglycemia-enhanced hemorrhagic transformation in a focal ischemia rat model. Neuroscience 169: 402-414.

71   Sato Y, Kajiyama S, Amano A, Kondo Y, Sasaki T, et al. (2008) Hydrogen-rich pure water prevents superoxide formation in brain slices of vitamin C-depleted SMP30/GNL knockout mice. Biochem Biophys Res Commun 375: 346-350.

72   Nagata K, Nakashima-Kamimura N, Mikami T, et al. (2009) Consumption of molecular hydrogen prevents the stress-induced impairments in hippocampus-dependent learning tasks during chronic physical restraint in mice. Neuropsychopharmacology 34: 501-508.

73   Ito M. et al. Drinking hydrogen water and intermittent hydrogen gas exposure, but not lactulose or continuous hydrogen gas exposure, prevent 6-hydorxydopamine-induced Parkinson's disease in rats. Med Gas Res. 2012;2(1):15. doi: 10.1186/2045-9912-2-15. [PMC free article]

74   Zhang J. et al. Effect of hydrogen gas on the survival rate of mice following global cerebral ischemia (Shock 37(6), 645–652, 2012) Shock. 2012;38(4):444. [PubMed]

75   Med Gas Res. 2018 Jan 22;7(4):247-255. doi: 10.4103/2045-9912.222448. eCollection 2017 Oct-Dec.
     Hydrogen-rich water for improvements of mood, anxiety, and autonomic nerve function in daily life.

76   Obesity (Silver Spring). 2011 Jul;19(7):1396-403. doi: 10.1038/oby.2011.6. Epub 2011 Feb 3.

77   Nutr Res. 2008 Mar;28(3):137-43. doi: 10.1016/j.nutres.2008.01.008.

78   J Diabetes Investig. 2017 Apr 8. doi: 10.1111/jdi.12674

79   www.docstoc.com/docs/24767241/Allergy-Effects-On-The-Pancreas-And-Small-Intestine/

80   Epithelial cells in pancreatic ducts are the source of the bicarbonate and water secreted by the pancreas. Bicarbonate is a base and critical to neutralizing the acid coming into the small intestine from the stomach. The mechanism underlying bicarbonate secretion is essentially the same as for acid secretion parietal cells and is dependent on the enzyme carbonic anhydrase. In pancreatic duct cells, the bicarbonate is secreted into the lumen of the duct and hence into pancreatic juice.

81   *The New Supernutrition*, Passwater, Richard A. Pocket Books, NY (May 1991).

82   Oharazawa H, Igarashi T, Yokota T, et al. Protection of the retina by rapid diffusion of hydrogen: administration of hydrogen-loaded eye drops in retinal ischemia-reperfusion injury. *Invest Ophthalmol Visual Sci.* 2010;51:487-492.

83   http://www.mgwater.com/marxneut.shtml

84   http://www.fastmag.info/sci_bkg.htm

85   *Journal of the American Academy of Nurse Practitioners.* December 2009, Volume 21, Issue 12, Pages: 651-657 "Oral magnesium supplementation in adults with coronary heart disease or coronary heart disease risk"

86   Saver JL, Kidwell C, Eckstein M, Starkman S; for the FAST-MAG pilot trial investigators. *Stroke.* 2004; 35: e106–108.

87   http://cat.inist.fr/?aModele=afficheN&cpsidt=21648128

88   Muir KW. Magnesium for neuroprotection in ischaemic stroke: rationale for use and evidence of effectiveness. CNS Drugs.2001; 15: 921–930.

89   Postgrad Med J 2002;78:641-645 doi:10.1136/pmj.78.925.641.

90   Resnick LM, Gupta RK, Gruenspan H, Alderman MH, Laragh JH: Hypertension and peripheral insulin resistance: possible mediating role of intracellular free magnesium. Am J Hypertens 3:373–379, 1990[Medline]

91   Ma J, Folsom AR, Melnick SL, Eckfeldt JH, Sharrett AR, Nabulsi AA, Hutchinson RG, Metcalf PA: Associations of serum and dietary magnesium with cardiovascular disease, hypertension, diabetes, insulin, and carotid wall thickness: the ARIC study. J Clin Epidemiol 48:927–940, 1985

92   Rosolova H, Mayer O Jr, Reaven GM: Insulin-mediated glucose disposal is decreased in normal subjects with relatively low plasma magnesium concentrations. Metabolism 49:418–420, 2000[Medline]

93   Med Gas Res. 2019 Jan 9;8(4):144-149. https://www.molecularhydrogenstudies.com/inhalation-of-hydrogen-in-parkinsons-disease/

94   http://www.ncbi.nlm.nih.gov/pubmedhealth/PMH0001762/

95   http://www.ncbi.nlm.nih.gov/pubmed/15885623

96   Hashimoto T, Nishi K, Nagasao J, Tsuji S, Oyanagi Kin Res. 2008 Mar 4;1197:143-51:

97   http://www.ncbi.nlm.nih.gov/pubmed/1...ubmed_RVDocSum

98   http://www.cannabis-med.org/english/bulletin/ww_en_db_cannabis_artikel.php?id=131#2

99   http://orthomolecular.org/resources/omns/v05n01.shtml

100   Clinical Toxicology, 2016; 54: 924-1109 http://www.drug-education.info/documents/iatrogenic.pdf

101   http://vaccinepapers.org/vaccine-aluminum-travels-to-the-brain/

102   Christian Science Monitor. June 3, 2005 – California takes aim at chemicals in plastics http://www.csmonitor.com/2005/0603/p02s01-uspo.html

103   EPA Journal – May 1985. Lewis, Jack.

104 https://www.nytimes.com/2009/09/28/health/policy/28vaccine.html

105 SOJ | Microbiology & Infectious Diseases. Hydrogen Medicine Therapy: An Effective and Promising Novel Treatment for Multiple Organ Dysfunction Syndrome (MODS) Induced by Influenza and Other Viral Infections Diseases?https://symbiosisonlinepublishing.com/microbiology-infectiousdiseases/microbiology-infectiousdiseases70.php

106 http://www.aap.org/advocacy/archives/mayautism.htm news release on a policy statement published in the May issue of Pediatrics, the peer-reviewed scientific journal of the American Academy of Pediatrics (AAP).

107 Vol. 114 No. 3 pp. 793-804 (doi:10.1542/peds.2004-0434)

108 http://www.ewg.org/reports/autism/part1.php

109 Curr Med Chem. 2005;12(10):1161-208. Metals, toxicity and oxidative stress.

110 Children with autism have mitochondrial dysfunction. University of California - Davis Health System

111 Molecular hydrogen increases resilience to stress in mice. Qiang Gao, Han Song, Xiao-ting Wang, Ying Liang, Yan-jie Xi, Yuan Gao, Qing-jun Guo, Tyler LeBaron, Yi-xiao Luo, Shuang-cheng Li, Xi Yin, Hai-shui Shi & Yu-xia Ma Scientific Reports 7, Article number: 9625 (2017) doi:10.1038/s41598-017-10362-6

112 The Clinical Application of Hydrogen as a Medical Treatment. Acta Med. Okayama, 2016 Vol. 70, No. 5, pp. 331-337. Okayama University Medical School.

113 Brain Research. Volume 1328, 30 April 2010, Pages 152-161

114 Hydrogen acts as a therapeutic antioxidant by selectively reducing cytotoxic oxygen radicals Ikuroh Ohsawa1, Masahiro Ishikawa1, Kumiko Takahashi1, Megumi Watanabe1,2, Kiyomi Nishimaki1, Kumi Yamagata1, Ken-ichiro Katsura2, Yasuo Katayama2, Sadamitsu Asoh1 & Shigeo Ohta1

115 volume 70, issue 12 (2007) of the Journal of Toxicology and Environmental Health, Part A; "oral antibiotics will reduce the amount of normal gut flora (which demethylate methylmercury) and may increase the amount of yeast and E. coli (which methylate inorganic mercury), resulting in both higher absorption and decreased excretion of mercury."

116 National Autistic Association news bulletin in response to Institute of Medicine Report Spring 2004 http://www.nationalautismassociation.org

117 Williams, Valeri. Vaccine preservative's effects may have been known. http://www.laleva.cc/choice/vaccine_preservative.html

118 Newsweek Magazine, July 31, 2000 and Care in Normal Birth: A Practical Guide—W.H.O's "Safe Motherhood" series

119 Mid Wife Info. "Immediate clamping of the umbilical cord can reduce the red blood cells an infant receives at birth by more than 50%, resulting in potential short-term and long-term neonatal problems." So concluded Judith Mercer, CNM and colleagues in a study reported in the fall of 2001

in the Journal of Midwifery and Women's Health (Mercer, 2001). "Early clamping of the umbilical cord at birth, a practice developed without adequate evidence, causes neonatal blood volume to vary 25% to 40%. Such a massive change occurs at no other time in one's life without serious consequences, even death. Early cord clamping may impede a successful transition and contribute to hypovolemic and hypoxic damage in vulnerable newborns" (Mercer, 2002).

http://www.midwifeinfo.com/feature-cordclamping.php

120   Morley, George M. Immediate clamping of the umbilical cord (ICC) at birth, a possible connection to Autism? ICC is routinely applied during premature, operative and "at risk" births, and increasingly during "normal" births following the recommendation (4) that a segment of the cord should be retrieved immediately after delivery for medico-legal purposes. The immediate effect of ICC is to deprive the neonate of placental respiration and transfusion resulting in complete asphyxia until the lungs function, and 30%-50% loss of the neonate's natural blood volume; the combined hypoxia and hypovolemia / ischemia is then conducive of hypoxic ischemic brain injury. The neonate that receives a full placental transfusion has enough iron to prevent anemia during the first year of life(5), but blood loss in a neonate subjected to ICC becomes evident in infancy as anemia.(5) In grade school children, anemia correlates with all types of autistic disorder (6) and the degree of anemia correlates with the degree of mental deficiency; (7) correcting the anemia does not correct the defect. Kinmond et al. (8) showed that delayed cord clamping combined with gravity assisted placental transfusion prevented anemia (the need for blood transfusion) in preemies. Hack et al. (9) found a high incidence of poor achievement in low birth weight babies. The correlation between autism and birth complications is supported by aother sudies. Hultman (10) reports a great increase in the risk of autism in cesarean deliveries, deliveries with fetal distress and five minute Apgar scores below seven. These obstetrical situations correlate with ICC.

121   Israels LG. Observations on vitamin K deficiency in the fetus and newborn: has nature made a mistake? Department of Medicine, University of Manitoba, Manitoba Institute of Cell Biology, Winnipeg, Canada. Semin Thromb Hemost 1995;21(4):357-63

122   http://www.sciencedaily.com/releases/2006/06/060618224104.htm

123   Ishibashi T, Sato B, Rikitake M, Seo T, Kurokawa R, Hara Y, et al. Consumption of water containing a high concentration of molecular hydrogen reduces oxidative stress and disease activity in patients with rheumatoid arthritis: an open-label pilot study. Med Gas Res. 2012;2:27.

124   Terawaki H, Hayashi Y, Zhu WJ, Matsuyama Y, Terada T, Kabayama S, et al. Transperitoneal administration of dissolved hydrogen for peritoneal dialysis

patients: a novel approach to suppress oxidative stress in the peritoneal cavity. Med Gas Res. 2013 Jul 1;3(1):14.

125 Nakayama M, Nakano H, Hamada H, Itami N, Nakazawa R, Ito S. A novel bioactive haemodialysis system using dissolved dihydrogen (H2) produced by water electrolysis: a clinical trial. Nephrol Dial Transplant. 2010;25:3026-33.

126 "Bad Cholesterol": A Myth and a Fraud; F. Batmanghelidj, M.D.; http://www.watercure.com/sci_myth.html

127 http://foodmatters.tv/articles-1/are-you-chronically-dehydrated

128 New hypothesis of chronic back pain: low pH promotes nerve ingrowth into damaged intervertebral disks C. LIANG et al;Acta Anaesthesiologica ScandinavicaArticle first published online: 7 MAR 2012; http://onlinelibrary.wiley.com/journal/10.1111/(ISSN)1399-6576

129 http://scialert.net/abstract/?doi=ajava.2012.420.426

130 iceberg lettuce 96%,squash, cooked. 90%,cantaloupe, raw, 90%
2% milk 89%,apple, raw 86%,cottage cheese 76%,potato, baked 75%
macaroni, cooked 66%,turkey, roasted 62%,steak, cooked 50%
cheese, cheddar 37%,bread, white 36%,peanuts, dry roasted 2%

131 http://www.diagnose-me.com/cond/C5223.html

132 http://www.watercure.com/

133 Kjaer A, Knigge U, Jørgensen H, Warberg J., "Dehydration-induced vasopressin secretion in humans: involvement of the histaminergic system." Am J Physiol Endocrinol Metab., 279.6 (2000):E1305-10.

134 The Lancet Oncology, news release, May 8, 2012

135 E. Tili, J.-J. Michaille, D. Wernicke, H. Alder, S. Costinean, S. Volinia, C. M. Croce. Mutator activity induced by microRNA-155 (miR-155) links inflammation and cancer. Proceedings of the National Academy of Sciences, 2011; 108 (12): 4908 DOI: 10.1073/pnas.1101795108

136 Powers SK, Jackson MJ. Exercise-induced oxidative stress: cellular mechanisms and imp

137 Ostojic SM. Serum alkalinization and hydrogen-rich water in healthy men. Mayo Clin Proc 2012; 87: 501–502

138 Pilot study: Effects of drinking hydrogen-rich water on muscle fatigue caused by acute exercise in elite athletes Kosuke Aoki, Atsunori Nakao, Takako Adachi,[1] Yasushi Matsui, and Shumpei Miyakawa

139 https://repositorio-aberto.up.pt/bitstream/10216/21741/2/39412.pdf

140 Med Gas Res. 2012; 2: 12. Pilot study: Effects of drinking hydrogen-rich water on muscle fatigue caused by acute exercise in elite athletes Kosuke Aoki, Atsunori Nakao, Takako Adachi, Yasushi Matsui, and Shumpei Miyakawa

141 Int J Sports Med 2015; Molecular Hydrogen in Sports Medicine: New Therapeutic Perspectives

142 https://www.ncbi.nlm.nih.gov/pubmed/22520831

143 http://www.sbrate.com.br/pdf/artigos/atualizacao_lesoes_musculares.pdf

144 http://www.tandfonline.com/doi/abs/10.3810/pgm.2014.09.2813

145 Mitochondria as an important target in heavy metal toxicity in rat hepatoma AS-30D cells;Belyaeva EA, Dymkowska D, Wieckowski MR, Wojtczak L.j; Toxicol Appl Pharmacol. 2008 Aug 15;231(1):34-42. Epub 2008 Apr 7. PubMed

146 Effect of mercury vapor exposure on metallothionein and glutathione s-transferase gene expression in the kidney of nonpregnant, pregnant, and neonatal rats;.Brambila E, Liu J, Morgan DL, Beliles RP, Waalkes MP; J Toxicol Environ Health A. 2002 Sep 13;65(17):1273-88. PubMed

147 Metal-mediated formation of free radicals causes various modifications to DNA bases, enhanced lipid peroxidation, and altered calcium and sulfhydryl homeostasis; PubMed

148 Free radicals, metals and antioxidants in oxidative stress-induced cancer. Valko M, Rhodes CJ, Moncol J, Izakovic M, Mazur M.; Chem Biol Interact. 2006 Mar 10;160(1):1-40. Epub 2006 Jan 23.;PubMed

149 Disorders of apoptosis may play a critical role in some of the most debilitating metal-induced afflictions including hepatotoxicity, renal toxicity, neurotoxicity, autoimmunity and carcinogenesis. Metals and apoptosis: recent developments.Rana SV. J Trace Elem Med Biol. 2008;22(4):262-84. Epub 2008 Oct 10; PubMed

150 Ohio State University Medical Center - Thu, 01/22/2009 - 16:05 http://www.diabetesincontrol.com/results.php?storyarticle=6461

151 In 1947, the British military discovered that oxygen could be toxic (Oxygen Toxicity) as a result of underwater research. Oxygen toxicity is a condition resulting from the harmful effects of breathing molecular oxygen at increased partial pressures. It is also known as oxygen toxicity syndrome, oxygen intoxication, and oxygen poisoning. Severe cases can result in cell damage and death, with effects most often seen in the central nervous system, lungs and eyes.

152 https://www.ncbi.nlm.nih.gov/pubmed/17656037

153 Acidity generated by the tumor microenvironment drives local Invasion; Veronica Estrella, Tingan Chen, Mark Lloyd, et al; Cancer Res Published OnlineFirst January 3, 2013; doi:10.1158/0008-5472.CAN-12-2796

154 Cancer-related inflammation, the seventh hallmark of cancer: links to genetic instability;Francesco Colotta1et al; Carcinogenesis vol.30 no.7 pp.1073–1081, 2009; Nerviano Medical Sciences, Nerviano, 20014 Nerviano, Milan, I; http://carcin.oxfordjournals.org/content/30/7/1073.full.pdf

155 UT Southwestern Medical Center. "Oxygen – key to most life – decelerates many cancer tumors when combined with radiation therapy."

ScienceDaily. ScienceDaily, 23 July 2013. <www.sciencedaily.com/ releases/2013/07/130723154959.htm>.

156 S. Thomas, M. Harding, S. C. Smith, J. B. Overdevest, M. D. Nitz, H. F. Frierson, S. A. Tomlins, G. Kristiansen, D. Theodorescu. CD24 is an effector of HIF-1 driven primary tumor growth and metastasis. Cancer Research, 2012; DOI:10.1158/0008-5472.CAN-11-3666; http://www.sciencedaily.com/ releases/2012/09/120913123516.htm

157 Kasper M.a. Rouschop, Twan Van Den Beucken, Ludwig Dubois, Hanneke Niessen, Johan Bussink, Kim Savelkouls, Tom Keulers, Hilda Mujcic, Willy Landuyt, Jan Willem Voncken, Philippe Lambin, Albert J. Van Der Kogel, Marianne Koritzinsky, and Bradly G. Wouters. The unfolded protein response protects human tumor cells during hypoxia through regulation of the autophagy genes MAP1LC3B and ATG5. Journal of Clinical Investigation, 2009; DOI: 10.1172/JCI40027

158 Maria Galluzzo, Selma Pennacchietti, Stefania Rosano, Paolo M. Comoglio and Paolo Michieli. Prevention of hypoxia by myoglobin expression in human tumor cells promotes differentiation and inhibits metastasis. Journal of Clinical Investigation, 2009; DOI: 10.1172/JCI36579

159 Burnham Institute. (2009, August 9). Unraveling How Cells Respond To Low Oxygen. ScienceDaily. Retrieved February 7, 2014 from www.sciencedaily. com/releases/2009/08/090805164915.htm

160 Measurements and combat of stress effects (author's transl); von Ardenne M.; ZFA. 1981;36(6):473-87; http://www.ncbi.nlm.nih.gov/pubmed/7336784

161 The pulmonary physician in critical care c 2: Oxygen delivery and consumption in the critically ill, R M Leach, D F Treacher. https://www.ucl. ac.uk/anaesthesia/StudentsandTrainees/OxygenDeliveryConsumption.pdf

162 https://www.ucl.ac.uk/anaesthesia/StudentsandTrainees/OxygenDelivery Consumption.pdf

163 J. Radiat. Res., 52, 545–556 (2011) How Can We Overcome Tumor Hypoxia in Radiation Therapy?

164 Impact of early sepsis on oxygen delivery in the microvasculature. Critical Care20059 (Suppl 1): P75

165 Influence of magnesium deficiency on the bioavailability and tissue distribution of iron in the rat. The Journal of Nutritional Biochemistry, Volume 11, Issue 2, Pages 103-108

166 http://bloodjournal.hematologylibrary.org/cgi/reprint/44/4/583.pdf

167 http://www.jbc.org/cgi/reprint/122/3/693.pdf

168 http://www.agclassroom.org/teen/ars_pdf/family/2004/05lack_energy.pdf

169 Terwilliger and Brown, 1993; Takenhiko and Weber; Wood and Dalgleish, 1973

170 http://bloodjournal.hematologylibrary.org/cgi/reprint/44/4/583.pdf

171 Ann Intern Med. 1971; 74 (4):632-633.

172 Int J Hyperthermia. 2010; 26(3): 232–246. Hypoxia-Driven Immuno suppression: A new reason to use thermal therapy in the treatment of cancer?

173 https://www.ucl.ac.uk/anaesthesia/StudentsandTrainees/OxygenDelivery Consumption.pdf

174 Alexander New; Emerg Med J. Feb 2006; 23(2): 144–146.;doi: 10.1136/ emj.2005.027458' http://www.ncbi.nlm.nih.gov/pmc/articles/PMC2564043/

175 http://en.wikipedia.org/wiki/Lactic_acidosis

176 Brewer, A. Keith PhD, Cancer, Its Nature and a Proposed Treatment, 1997; Brewer Science Library; http://www.mwt.net/~drbrewer/brew_art.htm

177 THE EFFECT OF pH ON THE OXYGEN CONSUMPTION OF TISSUES. Huntington Memorial Hospital Harvard University. (Received for publication, May 15, 1925.)

178 Undersea Hyperb Med. 2010 Jan-Feb;37(1):49-50. Sudden death during hyperbaric oxygen therapy: rare but it may occur. https://www.ncbi.nlm. nih.gov/pubmed/20369652#

179 http://www.ncbi.nlm.nih.gov/pubmed/10520649]

180 Henderson, Y. Carbon Dioxide. Article in Encyclopedia of Medicine. 1940

181 Immunologic Consequences of Hypoxia during Critical Illness. Harmke D. Kiers, M.D.; Gert-Jan Scheffer, M.D., Ph.D.; Johannes G. van der Hoeven, M.D., Ph.D.; Holger K. Eltzschig, M.D., Ph.D.; Peter Pickkers, M.D., Ph.D. http://anesthesiology.pubs.asahq.org/article.aspx?articleid=2524652

182 Int J Hyperthermia. 2010; 26(3): 232–246. Hypoxia-Driven Immunosuppression: A new reason to use thermal therapy in the treatment of cancer?

183 Front. Endocrinol., 23 October 2017 | https://doi.org/10.3389/fendo.2017.00279. The Emerging Facets of Non-Cancerous Warburg Effect

184 http://www.medicalnewstoday.com/articles/159225.php

185 BMJ. 1998 November 7; 317(7168): 1302–1306

186 Izv Akad Nauk Ser Biol. 1997 Mar-Apr;(2):204-17. Carbon dioxide–a universal inhibitor of the generation of active oxygen forms by cells (deciphering one enigma of evolution). Article in Russian

187 J Cell Mol Med. 2011 Jun; 15(6): 1239–1253. Why is the partial oxygen pressure of human tissues a crucial parameter? Small molecules and hypoxia

188 ibid

189 Acute normobaric hypoxia stimulates erythropoietin release.
Mackenzie RW[1], Watt PW, Maxwell NS.; High Alt Med Biol.; 2008 Spring; 9(1):28-37. doi: 10.1089/ham.2008.1043; http://www.ncbi.nlm.nih.gov/ pubmed/18331218

190 Int J Hyperthermia. 2010; 26(3): 232–246. Hypoxia-Driven Immunosuppression: A new reason to use thermal therapy in the treatment of cancer?

191 http://raypeat.com/articles/aging/altitude-mortality.shtml

192 Nicotinamide adenine dinucleotide, abbreviated NAD^+, is a coenzyme found in all living cells. The compound is a dinucleotide, since it consists of two nucleotides joined through their phosphate groups. One nucleotide contains an adenine base and the other nicotinamide. In metabolism, NAD^+ is involved in redox reactions, carrying electrons from one reaction to another. The coenzyme is, therefore, found in two forms in cells: NAD^+ is an oxidizing agent – it accepts electrons from other molecules and becomes reduced. This reaction forms NADH, which can then be used as a reducing agent to donate electrons.

193 http://www.rsc.org/Education/Teachers/Resources/cfb/transport.htm

194 http://drsircus.com/world-news/climate/co2#_edn5

195 CO2 footbath therapy; http://www.co2bath.com/top.htm

196 University of Tennessee; Hyperbaric Oxygen; http://www.utcomchatt.org/subpage.php?pageId=838

197 Acute normobaric hypoxia stimulates erythropoietin release. Mackenzie RW[1], Watt PW, Maxwell NS.; High Alt Med Biol.; 2008 Spring; 9(1):28-37. doi: 10.1089/ham.2008.1043; http://www.ncbi.nlm.nih.gov/pubmed/18331218

198 Arq Bras Cardiol. 2009 Jun;92(6):423-9, 440-7, 457-63. Effect of diaphragmatic breathing on heart rate variability in ischemic heart disease with diabetes.

199 British Journal of Cancer (2012) 106, 241–242. doi:10.1038/bjc.2011.576 www.bjcancer.com. Treating cancer with amplitude-modulated electromagnetic fields: a potential paradigm shift, again?

200 Clinics (Sao Paulo). 2011;66(7):1237-45. The therapeutic effect of a pulsed electromagnetic field on the reproductive patterns of male Wistar rats exposed to a 2.45-GHz microwave field. Kumar S1, Kesari KK, Behari J. *The New Supernutrition*, Passwater, Richard A. Pocket Books, NY (May 1991).

Oharazawa H, Igarashi T, Yokota T, et al. Protection of the retina by rapid diffusion of hydrogen: administration of hydrogen-loaded eye drops in retinal ischemia-reperfusion injury. *Invest Ophthalmol Visual Sci.* 2010;51:487-492.

http://www.mgwater.com/marxneut.shtml

http://www.fastmag.info/sci_bkg.htm

Journal of the American Academy of Nurse Practitioners. December 2009, Volume 21, Issue 12, Pages: 651-657 "Oral magnesium supplementation in adults with coronary heart disease or coronary heart disease risk"

Saver JL, Kidwell C, Eckstein M, Starkman S; for the FAST-MAG pilot trial investigators. *Stroke.* 2004; 35: e106–108.

http://cat.inist.fr/?aModele=afficheN&cpsidt=21648128

Muir KW. Magnesium for neuroprotection in ischaemic stroke: rationale for use and evidence of effectiveness. CNS Drugs.2001; 15: 921–930.

Postgrad Med J 2002;78:641-645 doi:10.1136/pmj.78.925.641.

Med Gas Res. 2019 Jan 9;8(4):144-149. https://www.molecularhydrogenstudies.com/inhalation-of-hydrogen-in-parkinsons-disease/

http://www.ncbi.nlm.nih.gov/pubmedhealth/PMH0001762/

http://www.ncbi.nlm.nih.gov/pubmed/15885623

Hashimoto T, Nishi K, Nagasao J, Tsuji S, Oyanagi Kin Res. 2008 Mar 4;1197:143-51:

http://www.ncbi.nlm.nih.gov/pubmed/1...ubmed_RVDocSum

http://www.cannabis-med.org/english/bulletin/ww_en_db_cannabis_artikel.php?id=131#2

Resnick LM, Gupta RK, Gruenspan H, Alderman MH, Laragh JH: Hypertension and peripheral insulin resistance: possible mediating role of intracellular free magnesium. Am J Hypertens 3:373–379, 1990[Medline]

Ma J, Folsom AR, Melnick SL, Eckfeldt JH, Sharrett AR, Nabulsi AA, Hutchinson RG, Metcalf PA: Associations of serum and dietary magnesium with cardiovascular disease, hypertension, diabetes, insulin, and carotid wall thickness: the ARIC study. J Clin Epidemiol 48:927–940, 1985

Rosolova H, Mayer O Jr, Reaven GM: Insulin-mediated glucose disposal is decreased in normal subjects with relatively low plasma magnesium concentrations. Metabolism 49:418–420, 2000[Medline]

Printed in the United States
by Baker & Taylor Publisher Services